Despine and the Evolution of Psychology

Despine and the Evolution of Psychology

Historical and Medical Perspectives on Dissociative Disorders

Edited and Translated by
Joanne M. McKeown
and
Catherine G. Fine

palgrave
macmillan

DESPINE AND THE EVOLUTION OF PSYCHOLOGY

First published in 2008 by
PALGRAVE MACMILLAN®
in the United States—a division of St. Martin's Press LLC,
175 Fifth Avenue, New York, NY 10010.

Where this book is distributed in the UK, Europe and the rest of the world, this is by Palgrave Macmillan, a division of Macmillan Publishers Limited, registered in England, company number 785998, of Houndmills, Basingstoke, Hampshire RG21 6XS.

Palgrave Macmillan is the global academic imprint of the above companies and has companies and representatives throughout the world.

Palgrave® and Macmillan® are registered trademarks in the United States, the United Kingdom, Europe and other countries.

ISBN-13: 978–0–230–60867–2
ISBN-10: 0–230–60867–1

Library of Congress Cataloging-in-Publication Data

Despine, Charles-Humbert-Antoine, 1777–1852.
[De l'emploi du magnétisme animal et des eaux minérales dans le traitement des maladies nerveuses, suivi d'une observation très curieuse de guérison de névropathie. English]
Despine and the evolution of psychology : historical and medical perspectives on dissociative disorders : a critical edition of A study of the uses of animal magnetism and mineral waters in the treatment of disorders of the nervous system followed by a case of a highly unusual cure of neuropathy by Dr. Charles-Humbert-Antoine Despine / edited and translated by Joanne M. McKeown and Catherine G. Fine.
p.cm.
Includes bibliographical references and index.
ISBN 0–230–60867–1
1. Dissociative disorders—Treatment. 2. Animal magnetism—Therapeutic use. 3. Mineral waters—Therapeutic use. I. McKeown, Joanne M. II. Fine, Catherine G., 1950– III. Title.
[DNLM: 1. Dissociative Disorders—therapy. 2. Balneology. 3. Hypnosis. WM 173.6 D468d 2008a]

RC553.D5D4713 2008
616.85'2306—dc22 2008015074

A catalogue record of the book is available from the British Library.

Design by Newgen Imaging Systems (P) Ltd., Chennai, India.

First edition: December 2008

10 9 8 7 6 5 4 3 2 1

Printed in the United States of America.

A Note about the Text

This first English translation from the French is a critical edition of Dr. Charles-Humbert *Antoine* Despine's *A Study of the Uses of Animal Magnetism and Mineral Waters in the Treatment of Disorders of the Nervous System Followed by A Case of a Highly Unusual Cure of Neuropathy* (1840), a seminal work in the history of medicine and psychology. It narrates Despine's decision to use hypnosis and his successful treatment of the paralyzed Estelle, an eleven-year-old Swiss child. A landmark in the study of multiple personality disorder, the monograph records one of the first cases of psychotherapy and therapeutic alliance.

This translation is dedicated to the humanism and industriousness of the Despine family. Traceable as far back as twelfth-century France, the extensive Despine family has a rich history of service to humanity. Public and private archives safeguard the stories, only one of which is told in these pages.

Philippe Despine, direct descendant of the medical doctor whose accomplishments are recorded in this monograph, merits special mention. It was his love for family and his passion for learning that supported and galvanized our vision, making this annotated and illustrated publication possible.

Contents

Photos

Photo 1 Portrait of Antoine Despine holding a plan of the Baths in Aix-en-Savoie, currently hanging in the Medical Director's office at the Baths, probably painted in 1820 by Prosper Dunant.

Source: Photograph by J. McKeown.

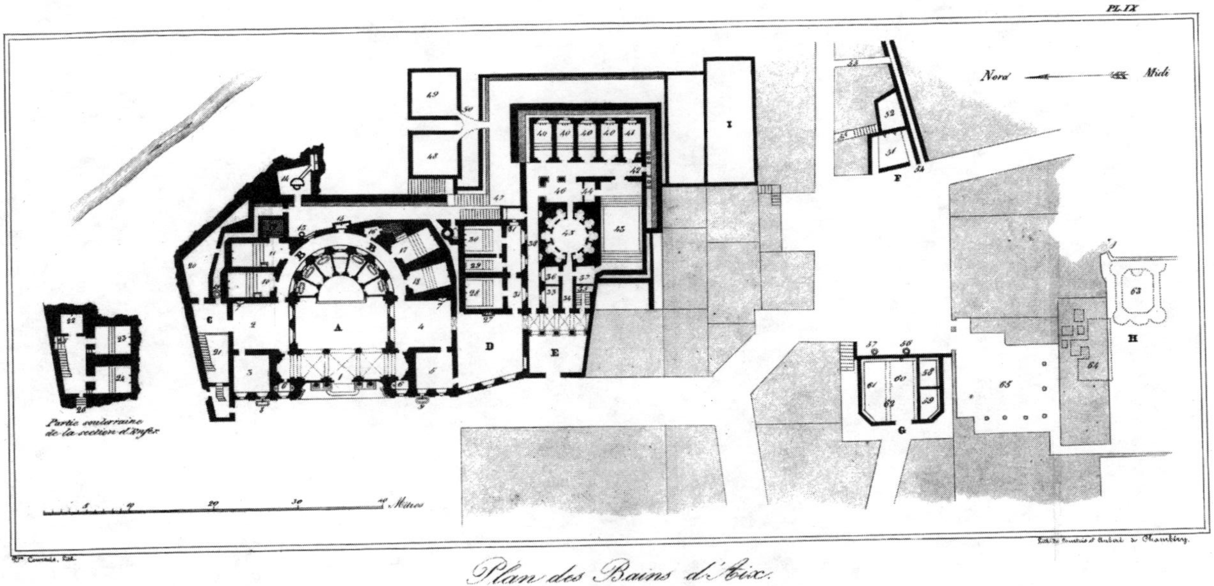

Photo 2 Map of the Baths in Aix-en-Savoie, originally published in *Manuel de l'étranger aux eaux d'Aix-en-Savoie* by Constant Despine (1834).

Source: From the personal library of Carole Koepke Brown.

Translators' Preface

Introduction

The first time Dr. Joanne M. McKeown met Antoine Despine (b. 2002), great-great-great-great-grandson of Antoine Despine (1777–1852), he was waiting to meet her in a child's seat in the back of his parents' car in a suburb of Dijon, France. The child's principal concern that afternoon (as his father had warned her) was that, due to a minor accident the day before involving Antoine's right hand, he might not be able to greet her properly with a hand shake. His father had reassured Antoine that *un bisou* would be equally acceptable, perhaps even preferable. When McKeown leaned into their car to say "*Bonjour!*" Antoine planted kisses on both cheeks. These were the first of many *bisous* she and Antoine exchanged during an unforgettable weekend in Burgundy with the Despine family.

McKeown is certain that the physician-author of *De L'Emploi du magnétisme animal et des eaux minérales dans le traitement des maladies nerveuses, suivi d'une observation très curieuse de guérison de névropathie* (A Study in the Use of Animal Magnetism and Mineral Waters in the Treatment of Disorders of the Nervous System Followed by a Case of a Highly Unusual Cure of Neuropathy) (1840) would have been proud of his young descendant, who not only endured a wound with resolve but welcomed a guest with a child's charming version of solicitude and chivalry. The nineteenth-century Antoine Despine himself showed these same qualities during his long careers as physician and Medical Director of the thermal baths in Aix-en-Savoie (now Aix-les-Bains). He was a dauntless, entrepreneurial man of science whose patients (and their families) occupied a significant part of his daily life and preoccupations. Despine wished to cure all who requested his care, regardless of social status and ability to pay for treatments. In addition, he was willing to use animal magnetism, an unconventional treatment modality (earlier called mesmerism, later known as hypnotism). Despine introduced this little understood and widely criticized therapy to treat nervous disorders resistant to other therapies. He allowed some patients to stay in his home so that he could better observe their behaviors and their response to treatment. Discouraged neither by his wife's and son's[1] complaints about this intrusion to their privacy nor by the professional criticism of animal magnetism, Despine acted in accordance with a vivid personal sense of responsibility to his patients, to science, and, ultimately, to humanity.

Despine's passionate, courageous, and broad-minded spirit permeates the 360-page pioneering work that has been the focus of McKeown's work for eleven years. She began translating the monograph in 1997 at the behest of a colleague in the English Department, Dr. Carole Koepke Brown. Familiar with Estelle's story from a summary of it in Henri Ellenberger's (1970) *Discovery of the Unconscious* and from Dr. Catherine G. Fine's 1988 article in the *American Journal of Clinical Hypnosis*, Brown needed a translation of the French monograph for a research project. After two sabbaticals, three trips abroad, and many hours of researching, rereading, and revising with Fine—who joined the team as Secondary Translator in 2001—and Brown as Volunteer Editor, the translated, annotated, illustrated monograph is complete.

THE TRANSLATION TEAM

McKeown's professional focus as a professor of French language, seventeenth and eighteenth century literary cultures, and of second language learning methodologies has broadened to accommodate work on this project. Her first publication (2006) related to Despine's successful treatment of Estelle L'Hardy appears in the *Journal of Christianity and Foreign Languages*. The paper explores parallels between the literary legacies of Despine and poet Clemens Brentano (1778–1842). It widens our understanding of the consequences of trauma as well as our understanding of inspiring visions in the midst of illness and physical and emotional pain (a second journal article is discussed later). Fine, born and educated in France, is an American-trained psychologist specializing in trauma therapy and hypnosis; she has treated dissociative patients in her clinical practice in Philadelphia for 25 years, using hypnosis as one treatment modality. In addition to her many publications, that include the often-cited article on Despine mentioned above, Fine also co-authored an article with Jean Goodwin about Despine's work that appeared as a book chapter in the series, *Clinical Practice* (1993). She was president of both the International Society of the Study of Multiple Personality Disorder (ISSMPD), currently renamed International Society for the Study of Trauma and Dissociation—ISST-D, and the American Society of Clinical Hypnosis (ASCH). Fine's medical knowledge, gained especially from her expansive clinical experience, enabled the team to make sense of behaviors and language characteristic of Estelle L'Hardy's disorder. McKeown and Brown—who acknowledge that Fine's invaluable time and experience were invested in the Despine project despite her already demanding commitments—are profoundly grateful. Brown, emerita associate professor of English, wrote "Pierre Janet's Addition to Charles-Humbert-*Antoine* Despine's 'old but very curious' Estelle L'Hardy monograph" (in press, *The Journal of Trauma and Dissociation*). In addition, she is involved in three local programs that protect at-risk children. All three contributors presented research papers in a symposium about Despine at the Conference for the International Society for the Study of Dissociation in 2002.[2]

ESTABLISHING THE TEXT

Brown established the text. Most commentators followed Ellenberger (1970) who had located only the widely disseminated (Paris, London, Lyon, Leipzig, Florence, Montpellier) 1840 reissue by Germer Baillière used for this translation. Burdet, however, had published the first issue in nearby Annecy, Despine's hometown, under the title *Observations de médecine pratique faites aux Bains d'Aix-en-Savoie* (*Observations in Clinical Medicine Made at the Baths of Aix-en-Savoie*). Although 1838 was printed as its year, the censor did not approve it and it was not licensed until October 5, 1839—as stated in both issues on p. 299 (Brown, in press). The Wellcome Institute for the History of Medicine (London Unit) kindly filmed a copy of the Burdet issue. Stephen Ferguson, curator of Rare Books at Princeton University Library, advised Brown on crucial matters in order to establish the text. With both Wellcome's 1838/9 copy and her own 1840 Baillière rare book open side by side, Brown identified few negligible differences. Since even the occasional odd spacing and typesetter's errors—such as an upside-down "u" font and a faulty font for "q"—usually appear in both issues, the Ballière was most likely printed from the Burdet standing type/plates. Thus, although the two issues are not identical, they are virtually the same except for their titles.

THE WORK OF TRANSLATING

The publication of Despine's work (1838/9 with its reissue in 1840) is one of trauma therapy's earliest monographs of a patient's recovery from a major dissociative disorder (formerly known as multiple personality disorder, now called dissociative identity disorder) using animal magnetism.

The translation team is certain that the industrious, ambitious Despine, convinced of the novelty and importance of his work with Estelle and other patients like her, rushed to publish and did not revise his work thoroughly (if at all); it clearly bears marks of a first draft. Despine used capitalizations, ellipses, and headings inconsistently, and he arbitrarily interchanged quotation marks and italics to indicate verbatim quotes. Italics also indicate medical terminology. At times, his use of pronouns was ambiguous. He was often repetitive and frequently used semicolons instead of periods, making run-on sentences commonplace. Other sentences are in fragments. Additionally, Despine rarely defined terms and there are inconsistencies, such as in dates.

Despine also made some incorrect assumptions about human physiology; indeed, he attempted to explain phenomena that were not well understood at the time. Among these were the way nerve networks function and the action of what he calls magnetic fluid on the body, including how the fluid acts like electricity. In these sections one gets the impression that Despine was using the act of writing to discern and form his own ideas rather than to recount what he had discovered. The prose here—formative and developmental rather than descriptive—lacks clarity, syntactical precision, grammatical accuracy, and coherence. It reveals a researcher who did not yet have mastery of his ideas. These sections typically raise more questions than they answer.

The organization of Despine's 360 pages is confusing, and at times renders understanding incomplete and fragmented, at best. For example, Despine used the term *époques* (periods/eras/ages/epochs) both to identify three major periods into which he divided the process of Estelle's healing, and to identify the phases of the magnetic cure, one of those three major periods. Since he used the same French word to speak of periods and the phases within a period, readers of the original can easily become confused about the chronology of treatment. Madame L'Hardy's transcriptions of the magnetic sessions (25 pages) that Despine himself described in the main narrative are found in an endnote. The separation of these closely related accounts creates fragments and overly complicates the presentation of this material. As a result, the reader's ability to understand and appreciate what occurs during these remarkable episodes is compromised. Additionally, some topics resurface intermittently as if for the first time; these topics include the use of metals in treatment, or "formulas" (now called algorithms in the broad sense of the word) used by patients to enter into a crisis (or hypnotic) state. We provide annotations to increase understanding, but the intermittent appearance of these subjects has made the placement of annotations tricky. Moreover, there is a *Répertoire Général* at the end of the monograph that briefly summarizes

30 notes. Despine wrote that notes 16–30, not included in the 1838/9 publication, would appear later; they are not, however, in the 1840 reissue. Nevertheless, references to these notes remain in his running text. In the *Préface* Despine also described plans for other publications about "the fruits of his own personal experience and research…and noteworthy observations and experiences of colleagues in Aix or doctors at other thermal establishments in Savoie" (pp. xix–xx). These "numbered installments" never materialized.

All of this contributes to what Fine has called a hysterical quality to Despine's prose, characterized by overzealous expectations, exaggerated sentiment, and disorganized, fragmented verbal flourishes. Fine explains that his writing no doubt parallel processes his experience of the course of treatment with Estelle. In "Restoring Literary Wholeness to Antoine Despine's Fragmented Account of the Magnetic Cure of Estelle L'Hardy's Dissociative Disorder" (2007) McKeown shows how the translation team has acted as literary therapists for Despine's confused and fragmented account as we have endeavored to give it cohesion and clarity. In fact, our goals have been (1) to produce a completely accurate rendering of the original, (2) to preserve its historical quality, and (3) to express the content clearly and coherently for a twenty-first century medical audience.

To achieve these goals we have honored Despine's exact mode of expression when doing so did not hinder understanding. If the meaning of a word in its original context is unclear, such as *vaporeux* and *la vie végétative*, we have used a cognate whenever possible and left its interpretation to the reader. If a cognate does not exist for that word or expression, we have provided a literal translation. The same principle guided our translation of whole sentences and even paragraphs that were unclear in the French. Since the original meaning was confused, the team translated for literal accuracy rather than for clarity; being perfectly clear would entail undue guessing and possibly misrepresenting Despine's knowledge. On the other hand, when Despine's meaning is clear but inadequately elaborated, the team either slightly revised the section or else included an annotation. In fact, we have fully annotated the work to establish an historical context and to provide a nineteenth-century understanding of electricity, electrotherapy, animal magnetism, and the use of metals in treating patients. Furthermore, we have corrected dates if we could discern Despine's intention, shortened sentences, and provided a glossary to define terms. A superscript (g) appears next to all glossed words. We have converted temperatures reported in Réamur to Celsius, leagues to kilometers, inches (*pouces*) to centimeters, and have used small capitals for Despine's emphasis.

To address organizational problems we have relabeled certain sections and adjusted their sequence. The order of some of Despine's footnotes has been changed; all of them appear here after the translation with our annotations but are preceded by the indicator, *Despine's footnote.* Madame L'Hardy's single footnote is handled similarly. We have inserted Despine's endnote 2 into the running text so that Madame L'Hardy's transcriptions of the magnetic sessions appear in their chronological location. Despine's remarks about work he never published have been omitted. To meet publishing requirements, the team cut sections of his Preface and Introduction, and summarized endnotes 7, 14, and parts of long endnote 15: reports on six patients in the First Series as well as all of the Second Series and Third Series. These summaries by Brown (boxed and italicized here) compromise neither the intent of the original nor the coherence and clarity of the translation. In addition, boxed italicized text indicates the translators' explanations for sections that need additional clarification.

For our wide-ranging readership—historians of psychology, historians of medicine, historians of culture, trauma therapists as well as trauma academicians—we decided on two indexes (compiled by Brown), one featuring names and the other, subjects. The information in the indexes is restricted to the monograph and the Notes and Glossary that follow it. Because Despine's scattered material on the same subject is often challenging to locate, it seemed appropriate to be generous in listing page numbers.

INDEBTEDNESS

Special mention must first be made of Carole Koepke Brown's contributions. To do them justice, however, is impossible. Not only did the idea to translate the entire monograph originate with Brown, but also the coherence, order, and clarity of the current publication are in great part her work. Brown's exceptionally close, pointed knowledge of Despine's study, in-depth research, and attention to detail in written expression enabled her to rework our early drafts in such a way that the English translation honors the author while appealing to current readership. McKeown and Fine are immeasureably grateful for her expertise, resourcefulness, and tenacity without which the project most certainly would not have been completed.

* * *

Our translation team is deeply indebted to the Despine family for its long-term, unwavering, and comprehensive support of this project. In particular, we are grateful to Philippe, the great-great-great-grandson of Antoine Despine and father of young Antoine, remembered fondly at the beginning of these remarks. *Docteur ès philosophie,* Philippe has written a biography of the Despine family (2005), whose roots date back to twelfth-century France. His *Biographie d'Antoine Despine: 1777–1852* has been cited numerous times in our annotations. Philippe gave us complete freedom with respect to the translation and its publication. As important, he gave McKeown unlimited access to materials in private archives, in the *Archives des Thermes Nationaux d'Aix-les-Bains,* in the *Archives municipales d'Aix-les-Bains*, and in the *Archives Départementales de la Haute-Savoie.* He has helped in other invaluable ways. Indeed, his benevolence seems to know no bounds.

McKeown was especially fortunate in Aix-les-Bains, where three individuals came to her assistance. Primary among them is Bernard Graber-Duvernay, M.D. When she began her research, he held Antoine Despine's position as Medical Director of the thermal establishment in Aix-les-Bains. Beginning on the hot June day in 1999 when McKeown met Bernard, cooled only by a tiny desk fan in his large office at the baths, he made it possible for her to work at Despine's library in Saint Innocent and at the *Bibliothèque Nationale de Médecine* in Paris. He also introduced McKeown to Geneviève Frieh-Giraud, historian of the Baths, and Joel Lagrange, archivist at the *Archives Municipales d'Aix-les-Bains.* Both provided nineteenth-century visuals, some of which are reproduced here, as well as relevant documents belonging to the Despine family. Although they knew McKeown only by association, Frieh-Giraud and Lagrange met McKeown questions and requests for access to documents and other materials with generosity and trust.[3]

McKeown three research trips were, in part, funded by grants from the *Faculty Development and Research Committee* at Moravian College. The College further supported her with two sabbaticals devoted to this work, and two *SOAR* (*Student Opportunities for Academic Research*) grants for senior French majors Kimberly Mabry and Lauren Anderson. Mabry (spring 2006) worked as primary translator of Philippe Despine's biography of Antoine Despine. Anderson (fall 2006) worked as primary translator of Philippe Despine's detailed footnotes and as editor of Mabry's original draft of the biography. Excerpts of Mabry's and Anderson's work appear in our annotations. Barbara Charvatova, exchange student from the Czech Republic, aided in the organization of our annotations. Bonnie Falla, M.L.S., and Dorothy Glew, Ph.D., public service and research librarians at Moravian College's Reeves Library, gave invaluable assistance. Likewise, Debbie Gaspar and Nancy Stroebel, public services librarians in circulation, hunted down many publications for the translation team. Also at Moravian, we are grateful for the support of then Dean Curt Keim, Ph.D., and faculty member Karen Keim, Ph.D., support that proved to be both practical and inspirational. Christie Jacobsen, Web master of extraordinary talent, used her considerable graphic design skills to format in QuarkXPress a spiral-bound packet of information (featuring nineteenth-century visuals in color) for prospective publishers. Brown served as an eager apprentice for Christie. The Center for Information Technology (CIT) staff members came to our aid countless times. Mickey Ortiz, Office Coordinator of College Central Services, did the complex work of assembling our documents for potential publishers. Secretary Robin Thomas located rare materials, created indices for rare books, and researched some questions for our team.

Other valuable assistance came from the knowledgeable, generous editor Ann West who consulted for the team and alerted us to The Bakken: A Library and Museum of the History of Electricity, located in Minneapolis, Minnesota. There, Elizabeth Ihrig, librarian, suggested that we put our most challenging questions to Dr. Manuel Parra of Santiago, Chile, who had done research at The Bakken. Indeed, he kindly contributed his expertise to our project.

Friends and family who have provided emotional and moral support for the duration of this long-lived project could never be adequately thanked. McKeown's stepdaughter Joy Rudi, now an elegant, independent 18-year old, was only seven when she helped McKeown organize binders where photocopies of the original monograph matched up with printouts of early drafts. McKeown's parents, four siblings and their families, and her dearest friends provided steadfast, unconditional support throughout. Catherine G. Fine would like to give special thanks to her husband, Robert Fine, M.D., her daughter, Stephanie G. Fine, M.Ed., her friends and colleagues, Richard P. Kluft, M.D., Nancy E. Madden, Psy.D. and Linda M. Young, M.D. and finally, her Mother, Ginette E. Baum, trilingual translator, for their timeless patience, generous understanding and willingness to forgo family and social activities in the service of this book.

The team has lived perhaps a bit like Despine, whose travels to increase his knowledge base and whose care of patients on weekends and holidays, sometimes at home and often for long hours at the baths, depended on backing from family and friends—even if it was at times given grudgingly. Each of us feels a certain poetic consolation in the similarity, and concedes, perhaps more willingly than the strong-willed, single-minded Despine would have, that had any of this support been lacking, the project would not exist as it does today.

CONCLUSION

Antoine's expectation of living on through his work long after his death has, it would seem, come true. People linked to Despine by blood or by study of his monograph admire and seek to emulate his industriousness, his compassionate generosity, and his love of inquiry and learning. Now, Despine's work will be studied by English speaking therapists, other professionals who facilitate recovery from trauma, academicians and historians of psychology and medicine. Thus, it should widen intellectual understanding of the complexities and evolution of trauma treatment. Indeed, *De L'Emploi du magnétisme animal* provides a bridge to the concepts developed by Jean-Martin Charcot (1825–1893), an immediate predecessor of Sigmund Freud (1856–1939), and Pierre Janet (1859–1947), Freud's contemporary (on the Despine-Janet connection, see Brown, in press). The translation fills a gap in the English-language genealogy of modern psychology and brings richness to our current understanding of dissociative disorders and post-traumatic stress disorder (PTSD). Until now, only commentaries on and quotes from the complete work have been available in English.

This preface began with a memorable anecdote from McKeown 2006 summer research trip. It is not possible to include the countless other stories that bring to life the relationships she now shares with Philippe and Florence Despine and their son Antoine, and with Bernard and Marie-Odile Graber-Duvernay. But the personal relationships—as important as the professional assistance provided by these people—bears underscoring, for it is reminiscent of Antoine Despine's perhaps most endearing quality: his desire to help and befriend people in need. McKeown's friendships with the Despines and the Graber-Duvernays will—without question—continue despite the completion of this project. This personal treasure-trove of cherished associations, an unexpected outgrowth of professional inquiries, bears an eerie—but altogether gratifying and satisfying—parallel to the physician's compassionate relationships with patients, ties that often continued long after they left Aix with their families and returned home, cured.

JOANNE MCKEOWN, PH.D.
Primary Translator
with Catherine G. Fine, Ph. D.
Secondary Translator
and Carole Koepke Brown, Ph.D.
Volunteer Editor

NOTES

1. Antoine Despine and Suzanne-Péronne Révillod (1784–1862) wed on April 16, 1806; they had three daughters and five sons. The first child, born March 19, 1807 at Annecy, was Constant who became a physician (P. Despine, 2005).
2. The Symposium, called "Raising Estelle, Again: Past & Current," featured these three papers: "Empathic Engagement: A Cautionary Tale of a Physician" by C.K. Brown, "Antoine Despine: Repetition Compulsion and Repetition Revulsion" by C.G. Fine, and "Antoine Despine's Magnetic Cure: Issues of Readership Yesterday and Today" by J.M. McKeown.
3. Geneviève Frieh-Giraud has written a beautiful and comprehensive coffee-table book (2005) about the history and current status of the thermal baths in Aix-les-Bains: *Les Thermes D'Aix-les-Bains: Le Fil de l'eau.*

WORKS CITED

Brown, C.K. (in press). Pierre Janet's addition to Charles-Humbert-*Antoine* Despine's "old but very curious" Estelle L'Hardy monograph. *The Journal of Trauma and Dissociation.*

Despine, Antoine. (1840). *De L'Emploi du magnétisme animal et des eaux minérales dans le traitement des maladies nerveuses, suivi d'une observation très curieuse de guérison de névropathie.* Paris: Germer Baillière.

Despine, P. (2005, unpublished manuscript). *La biographie d'Antoine Despine: 1777–1852.* Translator of text: Kimberly Mabry. Translator of footnotes and editor: Lauren Anderson.

Ellenberger, Henri F. (1970). *The discovery of the unconscious: The history and evolution of dynamic psychiatry.* New York: Basic Books.

Fine, C.G. (1988). The work of Antoine Despine: The first scientific report on the diagnosis of multiple personality disorder. *American Journal of Clinical Hypnosis,* 31, 32–39.

Fine, C.G. with J. Goodwin (1991). Mary Reynolds and Estelle: Somatic symptoms and unacknowledged trauma. In J. Goodwin (Ed.), *Trauma: Historical casebook of problem patients.* Washington, DC: American Psychiatric Press.

Frieh-Giraud, Geneviève (2005). *Les Thermes D'Aix-les-Bains: Le Fil de l'eau.* Spain: FIGEP'partner and Artes Graficas Toledo (groupe Mondadori).

McKeown, J.M. (2006). Visions as illness and inspiration: Young Estelle L'Hardy and Sister Anne-Catherine Emmerich in works of Doctor Antoine Despine and Poet Clemens Brentano. *Journal of Christianity and Foreign Languages,* 7, 29–43.

——— (2007). Restoring literary wholeness to the fragmented account of Antoine Despine's magnetic cure of Estelle L'Hardy's dissociative disorder. *International Journal of Clinical and Experimental Hypnosis,* 55 (4), 486–496.

Antoine Despine: Magnetizer and Pioneer in the Contemporary Treatment of Dissociative Disorders

Catherine G. Fine, Ph.D.

Animal magnetism, initially described in his *Dissertatio physico-medica de planetarum influxu* (1766) by Franz Anton Mesmer (1734–1815), has had since its inception a fluctuating, but recurring impact on the practices of medicine and psychology. Animal magnetism flourished after 1790 under the promotional influence of the Marquis de Puységur (1751–1825); he was fascinated by the unusual state of consciousness in which patients were very suggestible, able to diagnose their own illnesses, and prescribe effective remedies for their perceived conditions. Puységur and his contemporaries such as Pétetin (1785) noted the similarity between somnambulism in its natural state and magnetic sleep. The terms magnetic somnambulism or artificial somnambulism—ultimately relabeled hypnosis by James Braid (1795–1860)—evolved to capture the array of phenomena, curative, curious and sought, that would help explicate ailments that defied the conventional medicine of the day. Contemporary therapists would consider these conditions the venue of psychosomatic medicine. It is imbued with this spirit of exploration, inquisitiveness and pride that Antoine Despine (1777–1852), a provincial doctor, practiced medicine and treated other physicians' treatment failures with cutting edge interventions.

The current chapter offers an orienting outline to the reader of the translated Despine monograph (this volume) because Despine's assumptive and intellectual context was so different from our own and because, though an excellent physician, Despine was a poor and confusing writer. He rapidly alternated between excessive detail and broad generalizations. Perhaps without realizing it, he paralleled the process of hypnotic trance induction and deepening, without the benefit of reorientation. The reader deserves the assistance of a conceptual and organizational map of Despine's work because the anticipated psychotherapeutic anchors, such as explorations of the trauma genesis, the role of boundaries and transference-countertransference exchanges, are absent. Therefore, this chapter will attempt not only to capture the personal and professional influences on Despine's thinking, but it will also attempt to simplify and more clearly organize his presentation of facts and thus make it more accessible to the contemporary clinician.

Despine: Physician and Magnetizer

Despine's curative arsenal was not conventional for the time (Graber-Duvernay, 2005) and included hydrotherapy, electricity, and magnetism. Despine adopted the teachings of Barthez (1734–1806) (a former professor of his at the University of Montpellier) on vitalism in which some life force is presumed to unite the mind (soul) and the body. Despine's belief in vitalism imbricated easily with his evolving understanding of the new discoveries in physics and electromagnetism. Vitalism allows for the magnetic (hypnotic) phenomena scientifically observable in his cataleptic patients and did not offend his well-established Catholic faith—faith that he preserved and protected (McKeown, 2006).

The question that repeatedly begged to be asked was whether Despine was a lunatic or a luminary; there were proponents of both views. For some of his colleagues of the day, he was the former. A man duped by his patients and following mystical rather than scientific injunctions. For his gratefully cured patients and his followers (Charcot, Janet—well described in Brown (in press), Prince, and twentieth- and twenty-first century dissociative disorder specialists), he was the first physician who scientifically and systematically documented (Georget in Graber-Duvernay, 2005) and published on the diagnosis and treatment of a child with Multiple Personality Disorder (also called Dissociative Identity Disorder), an extreme dissociative condition (Kluft, 1984a).

Current readers of the Despine monograph (Despine, 1840; this volume) have the magical benefit of travel across time, continents, cultures, languages and medicopsychological formulations. They can be transported into the hypnogogic world of Despine, an educated physician-trailblazer who relished the science of steady, repeated observation prior to hypothesis formulation. Thus informed, he would co-lead (with his patients) the path to directed, flexible and often successful therapeutic interventions. With Estelle L'Hardy as his emblematic patient, Despine explored the systematic treatment of curious neuropathic symptomatology from the inside out.

Unlike current dissociative disorder specialists, Despine had no external treatment structure to instruct his interventions; he had no guidelines, no map of the psychophysical territory, no initial consideration that Estelle's presentation could be anything other than medical. He knew not the language of trauma, did not make a connection between Estelle's complex and perplexing symptoms and their potential post-traumatic sequellae. Over and above his scientific rigueur, Despine's commitment was to helping a distressed family and an ailing child. He was imbued with a sense of responsibility to his fellow man, of compassion for humankind, of faith in God and of duty to the Hippocratic Oath. He seems larger than life—a human in the service of mankind—a man gripped by a Promethean Will (P. Despine, 2005).

Yet, he was also a dreamer—a creative spirit, a decided, opinionated and determined experimentalist. He was a conflicted man, with strengths and faults. His delightful narcissism led him to the one mindedness of ingenious, circumvoluted interventions that track the patients' symptoms and requests. These qualities which invigorated some of his conclusions, also lead him to noteworthy and naïve self-deception. "Because of her age and youthful frankness, she [Estelle] cannot be suspected of emotionality, conceit, or the wish to cause a sensation. Finally, her extreme innocence made her incapable of deliberate deception" (this volume).

As a passionate scientist, Despine was so consumed by his work and his desire to observe repeatedly, completely and tie up loose ends that it left little time and room for his own family life. It did not help his personal life that, like physicians of the day, Despine moved his patients who traveled from afar into his own home making him imminently available for care, observation and unfortunately over-involvement. For instance, Despine described his important work with Estelle (this volume) taking place in the middle of the night and actually on Christmas day to accommodate the patient's perceived needs. His family, though no doubt at least somewhat accommodating (McKeown, 2007), had to adjust to his long and often unpredictable clinical hours, his unavailability during his time of exploratory readings, during his writing of his 1838/9 monograph and during his devoted letter writing to his colleagues and former patients in the service of his interest in hypnosis. His availability to his patients and to his personal exploratory endeavors rests not only on his personal motivation but also on a relatively solid familial foundation on which Despine could rely.

ESTELLE L'HARDY: CLINICAL CASE, EMBLEMATIC PATIENT

Estelle was born in Paris in 1825. Despine reported on a number of traumatic events that befell the girl and her family prior to Estelle's symptom onset (Fine, 1988; Fine 2003; this volume). He did not connect these traumas either to the fact or to the form of her symptoms. Starting at age five, Estelle was exposed to two major life-threatening epidemics in Paris; one was a measles epidemic when she was five, the other was a cholera epidemic when she was seven. Cholera killed her father and seriously compromised the life of her mother and sister. Both concurrent with and subsequent to these two life-imperiling events, Estelle developed high fevers, "head symptoms" that included severe headaches. In addition, the family described in Estelle an increasing hyperawareness, hyperalertness, and extreme hypersensitivity—symptoms consonant with a current diagnosis of post-traumatic stress disorder (PTSD). When she was nine, a simple fall became for Estelle, the opportunity to develop (or to let emerge) a variety of new symptoms that would increasingly immobilize her physically and paralyze her psychologically. At this point, current therapists would consider the opportunistic nature of dissociation and wonder whether this minor event (a fall that Estelle experienced when playing with a girl friend) could become a focus for subsequent somatization. All of her symptoms would rally her worried Mother's attention, significantly impede Estelle's affective, social, emotional and cognitive development, would interfere with normal separation individuation as well as other childhood developmental tasks. They would keep Estelle in a dependent, "safe" situation as Mother's little girl. Estelle's symptoms became increasingly compelling; her headaches augmented in both frequency and intensity. A nervous cough, oppression in the chest, pains in the stomach and rib cage and skin neuralgia became her constant and unrelenting companions; she always complained of being cold and insisted on eating a bizarre diet. Soon she became totally unable to walk on her own; the regression was almost complete. Estelle and her Mother consulted reputable physicians far and wide without success. Despine was their last hope.

Estelle initially reported to Despine hallucinations, illusions and disturbing dreams. Her waking hours were riddled with panic attacks.

Despine shifted from his initial therapeutic interventions which were more traditionally medical to magnetic somnambulism (hypnosis) when he suspected that Estelle was experiencing symptoms of ecstasy (trance phenomena) elucidated thanks to a conversation with Estelle's Mother about Estelle's disremembered conversations. In the magnetic sessions, Estelle revealed the presence of personalities/ego-states whom she identified initially as angels (called Angeline, Zéalida, Pansia, Elotina), though others emerged as the sessions proceeded (The Skeleton, Henriette, Athalie, men who grimaced, people in domestic altercations and Mademoiselle). Despine made note of the presence of these personalities, but seemed frankly disinterested in them other than in their relevance to Estelle. Was he parallel processing Estelle's *belle indifference* attitude about the personalities or was it that he himself was hypnotized/enthralled by the "little girl from Neuchâtel?" We will never know.

In his true observer nature, Despine seemed more focused on Estelle's behaviors, both in and out of session, as well as on the quality of her trance states than on the actual content of what she reported. The entire verbal exchange between Despine and Estelle was documented in the written notes of the Mother, present at all the treatment sessions (this volume). He centered his attention on the hypnotic process (since the cure was in the trance quality) rather than in the content of the verbalizations. The turning point in Estelle's therapy occurred when in crisis, she was able to get angry; subsequent abreactions ensued. Finally, Estelle predicted an "explosion of a hollow ball in her throat" (*globus hystericus* of somatic memory?); this release was accompanied by great somatic pain (in an undisclosed area), but from that point on, she could resume walking on her own. She rapidly regained her strength and returned to Switzerland June 30, 1837.

DISSOCIATIVE CONDITIONS THEN AND NOW: HOW WOULD DESPINE AND CONTEMPORARY PSYCHOTRAUMATOLOGISTS DISCUSS ESTELLE?

Signs and Symptoms of Dissociative Conditions

How would the contemporary well-trained trauma therapists view Estelle if she entered their twenty-first century practice? Current therapists would have a high index of suspicion prior to the first visit of the "Spoiled Child of Neuchâtel" from the notes of the previous physician, Dr. De Castella; they would anticipate in Estelle a dissociative disorder because of her reported polysymptomatic presentation spanning psychological and somatic domains, the protracted duration of her illness, and her numerous treatment failures at the hands of competent doctors. Contemporary therapists benefit from understanding that the active process of dissociation in which Behavior, Affect, Somatic/Sensation and Knowledge (BASK) (Braun, 1988) that would normally be associated was actively kept separate (Spiegel, 2001). They would also anticipate that alternate identities could form surrounding aspects of the dissociated material and suggest a history of significant traumatization (Putnam et al., 1986). In assessing the severity of dissociation, they would find it important to determine how disruptions of identity, consciousness or memory impede the achievement of Estelle's normal developmental tasks.

What would immediately leap to the forefront of the modern therapists' attention would be the comorbid prevalence of somatic and post-traumatic symptoms in the young Estelle. Her presentation justifies stringent inquiry about somatic symptoms (e.g., headache, stomach aches, and other undiagnosed pain) as well as somatoform dissociation because she too experienced loss of physical sensations, unusual pain tolerance and pain sensitivity as well as other sensori-perceptual anomalies (Nijenhuis et al., 1996). Despine reported that "Estelle had to spend the entire day in an absolutely horizontal position"; that she was "sensitive to temperature changes"; that "she could not hold up her head"; that "Her mother alone could move her without causing her terrible suffering"; that "She then would let out piercing screams"; that "The patient coughed intermittently, always in fits"; that "her pulse could not be taken without making her anxious and irritating her"; and, that "Estelle ate nothing until noon; she could eat neither meat nor broth. Constipation was customary" (this volume). Current therapists would also record the noteworthy post-traumatic symptoms that plague Estelle. These would include positive symptoms of PTSD, such as nightmares, night terrors, disturbing hypnogogic hallucinations, intrusive traumatic thoughts and memories, potential re-experiencing of flashbacks or traumatic reenactments. These would also include negative symptoms of PTSD, such as numbing and avoidance. Examples of PTSD symptoms from the monograph would be:

> While she slept, dreams, fantastic visions, and hallucinations reminded Estelle of painful memories from the previous evening or from her past and led her to periods of terror, fright, and constant dread—particularly

> when alone. A mouse, spider, fly, butterfly became (in her hypnogogic moments) robbers, monsters, or vampires. She believed they came to her at night with their terrible grimaces to devour or frighten her. (this volume)

With this plethora of PTSD and somatic symptoms, current therapists would look and listen for symptoms that are pathognomic of a dissociative disorder. They would watch for trance states with momentary slips of attention or nonresponsiveness. They would look to discover amnestic periods or moments of transient forgetting. Estelle experienced reported periods of self-absorption and times where she wanted to just be left alone. In her day-to-day life, she would lose time for recent activities in which she seemed fully participatory.

> It was common for Estelle to have a conversation or hear something of great interest to her, yet minutes later be unable to recall the smallest detail! Similarly, when we carried her on walks, she was interested in all she saw around her and chatted about it; yet, on her return Estelle often seemed amnestic for her experiences. . . . Estelle herself made a point of not speaking of it, despite her painful thoughts in this unusual state because she did not want to seem crazy or admit that she had no memory of many events. (this volume)

Current therapists would listen for the presence of alter personalities in adult patients or perhaps imaginary playmates in children. Estelle who was 11 years old when she encountered Despine was at the cusp of puberty when symptoms of both child and adult dissociativity ought be considered. Estelle's dissociative pathology showed an involvement in fantasy which interfered with normal activity; she perceived her "celestial guardians" to be present to not only keep her company and offer her solace but also to connect her with her adoring and adored defunct father.

> God, touched by my suffering and wanting to console and sustain me in this long trial, allows me to hear the singing of celestial spirits. . . . I have never been able to understand a single word, yet they speak to the heart, and I am very sure that Papa sings with them!! Yes, Momma, your Estelle would like to unite herself with these celestial spirits whose canticles so charm her. Besides, her state of suffering is so painful that she doesn't know how much longer she can endure!!! (this volume).

Noteworthy to the twenty-first century therapist would be Estelle's dysregulation in mood and affect as well as her identity alteration and state change as she spoke of herself in the third person. Estelle, like many dissociators, reported an overall pattern of anxiety, panic attacks, depression, rumination, and a subtle, to not so subtle, paranoia. Her depression was pervasive and unrelenting. "Profoundly affected by her perpetual suffering, Estelle had come to envision life and death with equal indifference. Estelle lived only because her death would hurt her wonderful mother" (this volume). As argued by Fine (1988), Estelle had a true and engrained dissociative disorder.

Treatment Goals in Somnambulists and Dissociators

Integrated functioning as a treatment goal is the consensus among current experts and was shared by Despine as well. This may require for the therapist to address the alternate identities as if they were separate but keeping in mind that it is only to bring about better communication and harmony among them. Despine understood the personalities as incidentals and more indicative of the complexity of the dissociative process than recapitulative of any trauma. This makes sense since Despine did not really recognize the post-traumatic nature of Estelle's condition. The current treatment guidelines for dissociative identity disorder (Chu et al., 2005) promote a triphasic treatment approach not unlike the treatment paradigm elaborated by Despine.

In stage one, the current therapist addresses patient safety, symptom stabilization and symptom reduction; in stage two, the therapist works directly with the traumatic memories; in stage three, the therapist expects identity integration and rehabilitation (Chu et al., 2005). Despine also divided his work with Estelle into three stages; in the First Stage of Despine's work with Estelle, he explored her various symptoms. He treated them with mineral waters, various baths, with electricity, with cauteries and with the medications/pharmacotherapy of the day. Most important in the First Stage, Despine was getting to know his patient, and intervening to deal with the diminution of her symptoms, just like current day therapists. The Second Stage that lasted only two intense days involved the psychoeducation of Madame L'Hardy and Estelle to the benefits of hypnosis for Estelle's recovery. Indeed, Despine had to convince the mother-daughter pair, not only by virtue of compelling arguments but also through providing for them published documents by master magnetizers, of the value of introducing magnetic somnambulism to Estelle's treatment. In the Last Stage (Phase One) of Despine's work, he introduced the induction of somnambulism (hypnosis) as a treatment modality. This led to the release of unacknowledged feelings and behaviors with further symptom reduction as one would expect in abreactions. In his Last Stage (Phase Two), the abreactions continued in milder form alternating between various dissociative states that, as time went on, became increasingly similar and eventually seemingly identical. Like with contemporary therapists, the Last Stage (Phase Three), here entrusted to the supervision or Madame L'Hardy because Estelle left Aix-en-Savoie, focused on unification and integration of the various self states.

Reflections on the Unfolding Treatment of the Dissociative Disorder Treatment: Then and Now

Estelle like other dissociative patients had predictable problems such as struggles with affect regulation, dissociative trances, and body image difficulties in part manifested in disordered eating. Estelle also exhibited self-destructive attacks on the body when she pulled at her hair and summarily cut it. From treatment onset, her primary presentation was one of major somatization. These culminated in complete catalepsis with bouts of continual and repetitive somatic crises followed by what Despine called magnetic crisis accompanied by changes in self-state. These switches that became increasingly overt as treatment progressed could explain why Estelle had many relational difficulties. She was often in the middle of some conflict in the household because of major problems in trust and enmeshment with her Mother. Estelle and her family were overwhelmed with the protracted nature of the young girl's cataleptic condition that was refractory to standard treatments of the day. The overprotective widowed mother was at a loss and perhaps colluded with Estelle in seeing the world as a dangerous place for her damaged daughter who carried both shame and responsibility for her illness as Herman (1993) and van der Kolk et al. (1996) would have predicted. The innovative organization of Despine's treatment of Estelle unfolded almost identically to what Kluft (1991) would urge as current and essential.

Kluft (1991) would recommend that, particularly in the initial stages of the therapy, the therapist ought to pay close attention to the solid establishment of the treatment frame with careful focus on the evolution of the therapeutic alliance. Indeed, Despine forged an empathic connection with the whole child; she ultimately felt comfortable sharing with him disowned experiences contained in voices, in dreams and in imaginary friends or self-states. Despine did not flinch in his steadfast devotion and acceptance of Estelle and all her experiences. He allowed for her various forms of expression, and thus perhaps facilitated her self-acceptance across all dissociative barriers including understanding the internal influences that led to her destructive or disruptive behaviors.

In the First Stage and Second Stage of Estelle's treatment by Despine that ranged from July 1836 to December 22, 1836, Despine listened to Estelle and listened well. He established with her a sturdy therapeutic alliance in which he restored some hope. He helped her feel safe and succeeded in diminishing some of her symptoms. Indeed, thanks to the numerous baths, showers, other forms of hydrotherapy, electrotherapy and the therapeutic uses metals, Despine had enabled Estelle to recover some movements in her formerly paralyzed legs. He helped her modulate some of her psychophysiological arousal as well as her affect. It was not so much that her fears, terrors, palpitations and nightmares were gone, but they seemed more tolerable and normative for her condition. In his care, Estelle became more self-contained and more focused on improving her physical limitations. She slowly regained control over her own body movements. In addition, during those months, Despine had become a steady, predictable figure for Estelle—a person who genuinely wished her well. He offered a focus for integration and for the development of a coherent self. He helped her to adapt better to her current environment rather than to expect the reverse.

Despine did not introduce hypnosis into the treatment of the young Estelle until the winter, five months after their work started. He actually only understood that hypnosis could be a valuable asset in the treatment of Estelle the day that Estelle was to leave Aix to return to Neuchâtel—before her planned return visit in the Spring of 1837. Indeed, after a conversation with Estelle's mother surrounding some bizarre experiences she had observed in her daughter, Despine concluded that

> Estelle had experienced phenomena that belonged to the pathological state nosologists call ecstasy. I had often observed it in my cataleptics. From that moment on I suspected more than ever that Estelle suffered from a functional ailment rather than an organic ailment. (this volume)

It is on this basis that at the end of December 1836, Despine, convinced that he could provide further help to Estelle through the use of magnetic passes, proposed to her Mother and to her the possibility of introducing magnetic somnambulism as a treatment modality. He presented it with great circumspection and with caution, but also legitimate hope. He reported to the family that his enthusiasm for using magnetism was incited by the works of Pétetin, Bertrand, and Foissac, works that he shared with them. The pivotal moment for Despine in his decision to use magnetism came when it was understood that Estelle was speaking out loud to a choir of angels.

Still not thinking in terms of multiple personality disorder or dissociative identity disorder, though perhaps thinking of hysteria (Despine, his endnote 4 in this volume), Despine did understand that Estelle had different (somnambulistic) states which guided her actions and feelings. Ultimately, an Inner Self Helper (ISH) named Angeline would emerge in the course of the therapy—that provided guidance in her life. This ISH would eventually inform her magnetic treatment. Despine delineated a crisis state and a noncrisis state, each with their own independent behaviors, affects, themes, and even physiologies. His goal was to access with Estelle these different states, learn about them, let them express themselves and see where this strategy would lead. Despine, by tracking his young patient astutely and with great acceptance, embodied the welcoming audience and listener which enabled Estelle in Phase Two of her treatment, to access and express feelings which had been to date unacceptable such as anger, rage, shame, helplessness, confusion.

During the Last Stage of Estelle's treatment, Despine introduced hypnosis to her therapeutic regimen. He recorded three phases within her magnetic cure: a Phase One of heterohypnosis, a Phase Two of spontaneous autohypnosis and a Phase Three best called post integration phase.

Phase One of the Magnetic Cure was amply described by both Despine and Estelle's Mother who took rigorous notes of both Despine's interventions and Estelle's responses to them. The Mother also recorded suggestions initiated by Estelle and her other selves. In this Phase One of the Magnetic Cure, Estelle seemed to travel through a number of differentiated ego-states/personalities and facilitated and explored the shifts among them.

Phase Two of the Magnetic Cure surrounded several incidents of "domestic disputes" in Estelle's presence; these combined with her newly accessed ego-states seem to have produced a greater disinhibition of Estelle's affect. She became increasingly angry. Despine spoke of Estelle "having fits of rage." With these fits came more shifting in and out of ego-states/personalities, as well as an augmented co-consciousness and blending amongst them. The crisis state had been the healthy state for Estelle. This Phase Two of the hypnotherapy, like in the work of contemporary dissociative disorder therapists, corresponded to the abreactive phase of trauma treatment where strong and disowned/disavowed affects were released and metabolized with ensuing improvements in somatic pathologies.

In Phase Three of the Magnetic Cure, Despine made no direct observations of the patient but relied on reports and letters made to him from divergent sources—people who could observe Estelle or who lived with her. This Phase Three occurred after Estelle left Aix to return to Switzerland, late in June 1837. What evolved most of the following year was an ebb and flow of Estelle's neuropathic symptoms, where crisis states alternated with noncrisis states at the will of Estelle. Predictably, psychosomatic symptoms from the past reemerged temporarily to submerge again. Each reincarnation of the experience seemed more muted than the previous one. All reported to be under Estelle's control.

Circumstances were such that Despine would never work with (or even see) Estelle directly again, but would just be informed of her continued progress and her adjustment to normal adolescent endeavors. He acknowledged even after this seeming treatment success that many factors contributed to Estelle's recovery—only one of which was the magnetic somnambulism. Despine's emblematic patient informed his treatment successes with other cataleptic men and women whose stories he also recounted in his monograph (this volume). Despine's advancing age and his multiple roles as clinician to a large demanding practice and as administrator of the Baths restricted his time and his energy needed to continue publishing all the information that he had accumulated on magnetic somnambulism. Therefore, his summary monograph remains invaluable.

Present-day traumatologists owe to Despine the initial medical forays and scientific explorations into the worlds of the somnambulists and cataleptics. He sorted through the numerous and often confusing clinical presentations of his subjects and targeted those symptoms that seemed most informational and significant for the patient. Like for Caul in Kluft (1984b), the symptoms presented by the patient dictated the interventions proposed by the magnetist, irrespective of the theoretical background of the therapist/magnetist. Despine delineated the stage-wise therapy structure to which we still subscribe (Fine, 1991; Herman, 1992; Kluft, 1991). He promoted the useful intercession of hypnosis in the treatment of dissociative disorders for symptom reduction, symptom exploration and symptom unfolding; some of these interventions, modified, are still current.

From the broader perspective of the emergent history of dissociative disorders, he anticipated the ebb and flow of popularity of these conditions and appreciated their extraordinary nature—nature that challenged the credibility of the patient and the therapist. He forewarned of the professional pitfalls threatening doctors who treated these disorders, yet also accurately anticipated the exhilaration of their successes. He urged young doctors to observe their patients carefully and often; he also encouraged them to listen well—for in this partially lies the treatment success.

Whether we invoke human nature, spiral dynamics, repetition compulsion or some other all too familiar explanation for the discovery and rediscovery of the facts (or the Truth as Despine would say), Despine pierced the veil of silence for the dissociative disorders. He challenged his contemporaries to think beyond the teachings of the past and include new ideas and new findings. And most important, Despine continues to challenge us neither to forget the past nor to ignore the present responses and truths of our patients. May we observe well and listen well.

Works Cited

Braun, B.G. (1988). The BASK model of dissociation. *Dissociation,* 1 (1), 4–23.

Brown, C.K. (in press). Pierre Janet's addition to Charles-Hubert-Antoine Despine's "old but very curious" Estelle L'Hardy monograph. *Journal of Trauma and Dissociation.*

Caul, D. (1984). Outcomes. In Kluft (1984b).

Chu, J.A., R. Loewenstein, P.F. Dell, P.M. Barach, E. Somer, R.P. Kluft, D.J. Gelinas, D.J. et al. (2005). Guidelines for treating dissociative identity disorder in adults. *Journal of Trauma and Dissociation*, 6 (4), 69–149.

Despine, C.H.A. (1840). *De L'emploi du magnetisme animal et des eaux minerales dans le traitement des maladies nerveuses, suivi d'une observation tres curieuse de guerison de nevropathie.* Paris: Germer Baillière.

Despine, P. (2005, unpublished manuscript). *La biographie d'Antoine Despine: 1777–1852.* Translator of text: Kimberly Mabry. Translator of footnotes and editor: Lauren Anderson.

——— (2006, unpublished manuscript) Commentary on "Le magnetisme au temps d'Antoine Despine." Graber-Duvernay (2005).

Fine, C.G. (1988). The work of Antoine Despine: The first scientific report on the diagnosis of multiple personality disorder. *American Journal of Clinical Hypnosis*, 31, 32–39.

Fine, C.G. (1991). Treatment stabilization and crisis prevention in the treatment of multiple personality disorder. *Psychiatric Clinics of North America*, 14, 661–675 (Guest Editor R. Loewenstein).

——— (2003, January). Repetition compulsion; repetition revulsion. In C.G. Fine (Chair), *Raising Estelle, again*. Symposium conducted at the annual meeting of the International Society for the Study of Dissociation, Baltimore, MD.

——— (in press). The vicissitudes of unacknowledged countertransferences: Antoine Despine's missed opportunities. *International Journal of Clinical and Experimental Hypnosis.*

Graber-Duvernay, B. (2005, unpublished manuscript). *Le magnetisme au temps d'Antoine Despine.*

Herman, J.L. (1992). *Trauma and Recovery: The aftermath of violence—from domestic abuse to political terror.* New York: Basic Books.

——— (1993). Sequelae of prolonged and repeated trauma: Evidence for a complex posttraumatic syndrome (DESNOS). In J.R.T. Davidson and E.B. Foa (Eds.), *Posttraumatic stress disorder: DSM-IV and beyond* (213–228). Washington, DC: American Psychiatric Press.

Kluft, R.P. (1984a). Multiple personality in childhood. *Psychiatric Clinics of North America,* 7, 121–134.

——— (1984b). Treatment of multiple personality disorder. *Psychiatric Clinics of North America,* 7, 9–29.

———(1991). Multiple personality disorder. In A. Tasman and S.M. Goldfinger (Eds.), *American Psychiatric Press annual review of Psychiatry* (Vol. 10, 161–188). Washington, DC: American Psychiatric Press.

McKeown, J.M. (2006). Visions as illness and inspiration: Young Estelle L'Hardy and Sister Anne-Catherine Emmerich in works of Doctor Antoine Despine and Poet Clemens Brentano. *Journal of Christianity and Foreign Languages*, 7, 29–43.

——— (2007). Restoring literary wholeness to the fragmented account of Antoine Depine's magnetic cure of Estelle L'Hardy's dissociative disorder. *International Journal of Clinical and Experimental Hypnosis*, 55 (4), 486–496.

Mesmer, F.A. (1766). *Dissertatio physico-medica de planetarum influxu.* Vienna: Ghelen.

Nijenhuis, E.R.S., P. Spinhoven, R. Van Dyck, O. Van der Hart, and J. Vanderlinden (1996). The development and the psychometric characteristics of the Somatoform Dissociation Questionnaire (SDQ20). *Journal of Nervous and Mental Disease*, 184, 688–694.

Pétetin. (1785). *Mémoire sur la découverte des phénomènes que présentent la catalepsie et le somnambulisme, symptômes de l'affection hystérique essentielle.* Lyon.

Putnam, F.W., J.J. Guroff, E.K. Silberman, L. Barban, and R.M. Post (1986). The clinical phenomenology of Multiple Personality Disorder: A review of 100 cases. *Journal of Clinical Psychiatry,* 47, 258–293.

Spiegel, D. (2001). Deconstructing the dissociative disorders: For whom the Dell tolls. *Journal of Trauma and Dissociation*, 2, 51–57.

Van der Kolk, B., D. Pelcovitz, S. Roth, F.S. Mandel, A. McFarlane, and J. L. Herman (1996). Dissociation, somatization, and affect dysregulation: The complexity of adaptation to trauma. *American Journal of Psychiatry*, 153 (suppl.), 83–93.

Appreciating Despine

Richard P. Kluft, M.D.

In the early 1970s I began to identify and treat patients who now would receive the diagnoses of Dissociative Identity Disorder or Dissociative Disorder Not Otherwise Specified. I was still a young psychiatrist in training. When I discussed the symptoms and stories of these patients with my teachers and mentors, I was greeted with skepticism, bewilderment, and downright disbelief. How could I have encountered several patients with a disorder so rare that most experienced clinicians had never recognized it in a single case? Furthermore, some of them believed that the condition might be purely iatrogenic. What egregious countertransference errors was I committing and what narcissistic grandiosity so compromised my character so that I could delude myself that I had seen such patients, and had induced them to enact such performances?

When *Sybil* was published (Schreiber, 1973), I hoped that her therapist, Cornelia Wilbur, M.D., might be a helpful resource for me. However, some of my prestigious professors insisted that Sybil was either iatrogenic or a fraud, and that Dr. Wilbur was a quack or worse. Allison's (1974) therapeutic approaches resembled exorcisms, and while some of his insights were useful, his overall approach did not seem consistent with several basic principles of psychotherapy as I understood it.

Feeling very much alone with the situation, I found myself returning over and over to the descriptions of phenomena and treatment of dissociative patients in Henri Ellenberger's (1970) classic history of the origins of dynamic psychiatry, *The Discovery of the Unconscious.* I was most impressed by Ellenberger's account of Despine's treatment of Estelle, the first description of a successful treatment of dissociative identity disorder. I was already immersed in the study of both psychoanalysis and hypnosis, and found it easy to recognize, even in Ellenberger's brief synopsis, that despite the limitations of his scientific knowledge, Despine implicitly was working with concepts I was being taught by my modern mentors and studying in my affiliation with psychotherapy researcher Lester Luborsky (e.g., Horvath and Luborsky, 1993; Luborsky et al., 2002). For example, instead of discouraging the appearance of other personalities, lest they be reinforced (the approach recommended to me by several professors), Despine engaged in respectful dialogs with them. It seemed to me that Despine, despite the fantastic theories of his era, had inadvertently discovered something crucial. By hearing out the various personalities and taking a flexible stance toward them, this pioneer was encouraging rather than suppressing the expression of important material and was forming a helping alliance with the various other personalities rather than attempting to suppress or rid the patient of them. This was consistent with the uncovering/exploring mission of psychoanalysis and the appreciation of the helping relationship or therapeutic alliance, a key aspect of Luborsky's research. The study of the other personalities and their behaviors often revealed that these phenomena reenacted past relationships and experiences long before the details of the patients' traumatic pasts could be discovered (Kluft, 2006).

Notwithstanding the passage of 130 years between Despine's publication of *A Study of the Uses of Animal Magnetism and Mineral Waters in the Treatment of Disorders of the Nervous System* and my own first contact with dissociative identity disorder in 1970–1971, Despine had (in modern parlance) what my professors lacked, *street creds*. He had been there. He had passed the test of healers from the prehistoric efforts of shamans to the most advanced high-tech intensive care units of today: the patient got well and went home!

I was able to obtain a copy of about 30 pages of Despine's manuscript, and worked out a crude translation. I developed an amiable, if at times conflicted relationship with Despine, in which I felt mentored by my study of critical aspects of his report, but unsettled by the rationales he used to understand what he observed and inform his interventions. I came to my first appreciation of what is recognized to be a key finding in the dissociative disorders field: that no matter what theories different effective clinicians used to explain what they did, they often wound up doing very much the same thing, suggesting that the realities of working effectively with dissociative disorder patients compel skillful clinicians to make similar interventions under similar circumstances, no matter how they account for what they have done (Caul, 1984; Kluft, 1984). As I changed my way of working with the various personalities to be more congruent with Despine's approach, I discovered both its promise and its pitfalls. One forms a more useful therapeutic relationship and enjoys more general cooperation than one experiences if his or her effort is to eliminate, suppress, or cast out these alters, but one enters a realm in which personalities may attempt to control, manipulate, and otherwise tyrannize the therapist, and in which enactments become an inevitable aspect of the therapeutic process, requiring zealous efforts to process in a manner that is beneficial to the treatment. These are the features of work with dissociative disorder patients that led many of Despine's contemporaries to see him both as self-deceived and as duped by his patients in general and by Estelle in particular.

This change in stance toward the personalities and my developing hypnotic interventions inspired both by Estelle's imagery and autohypnotic healing efforts and by some of Despine's therapeutic interventions enabled me to work more effectively with dissociative patients. Soon I could publish observations on my successful work with a large number of these patients (1982, 1984, 1986, 1993), and describe their response to a treatment approach that draws heavily upon the work of Despine and Janet (Kluft, 1991, 2006). These concepts informed long-running courses I directed at meetings of the American Psychiatric Association, the American Society of Clinical Hypnosis, and the International Society for the Study of Multiple Personality and Dissociation (which has since changed its name twice, first to the International Society for the Study of Dissociation, and then to the International Society for the Study of Trauma and Dissociation). Many modern contributors to the dissociative disorders field have either attended one of these courses or have been taught by those who did. In this largely indirect manner, Despine has been a major force in the resurgence of interest in the dissociative disorders, and is owed a debt of gratitude by clinicians, researchers, and dissociative disorder patients throughout the world. The moment a clinician moves to engage an alter personality in therapeutic dialog, he or she is walking in the footsteps of Despine.

How can we enter the world of Despine in order to understand what he is trying to say, and why he is trying to say it? It is difficult to contextualize, to look back and appreciate the circumstances, intellectual ambience, and overall situation of a physician over 170 years ago. My reading of Despine convinces me that within the pages of his monograph we encounter a complex emotional and scientific drama in media res, without sufficient back-story to make the overall project understandable. I have drawn extensively on materials provided me from Catherine G. Fine's research. Any errors of understanding or judgment based on those materials are my own. Here I can only hint at some of the factors that impress me. I cannot offer firm assurances of my accuracy.

Dr. Charles-Humbert-Antoine Despine was the director of the Baths at Aix-en-Savoie, following the footsteps of his father before him. He was working in a part of France then occupied by Italy, and is clearly working to conform some of his remarks to what is politically correct under the circumstances. A man in his 60s, he was aware that he was past his physical peak, doubting the strength of his remaining magnetic power. He venerated his father and hoped to be worthy of him. Despine's own son did not share his father's views, a cause of much personal distress. Despine was painfully aware that although many looked to him with respect and gratitude for his efforts, he was constantly under attack by colleagues who disapproved of him and his methods. (Similar polarization often surrounds those who work with the dissociative disorders today!)Despine is clearly driven to prove himself, his methods, and his theories. He believes that in Estelle, he has found the living paradigm of his scientific thinking. She would seem to demonstrate his diagnostic and therapeutic acumen, the worth of magnetism and the other modalities he employs, and the validation of vitalism. Much of the correspondence cited in his book is laudatory and confirmatory of both Despine and his perspectives.

Yet he has a particularly steep and uphill road to climb. Half a century previously, the Royal Commission had declared Mesmer's scientific explanations for his theory of animal magnetism unsound. Although the French Academy of Sciences had more recently taken a stance that declared magnetism was worthy of scientific study, it still was highly controversial and roundly attacked. Furthermore, Despine was a graduate of the medical school at Montpellier, which espoused a philosophy of vitalism; that is,that matters of body and spirit share importance both in illness and in health. This contrasted to the philosophy of the medical faculty of Paris, which was more organic and rationalistic, seeing the body of an ill person as a machine to be repaired by the appropriate interventions.

According to Kuhn (1970), author of the classic *The Structure of Scientific Revolutions*, those embracing alternative paradigms understand given observations to have different meanings consistent with their own paradigms. He observed (Kuhn, 1970) that proponents of different paradigms experience and understand the same things so differently that they, in effect, live in different worlds. What was scientific, reliable, and valid to a vitalist (Montpellier paradigm) might appear to be the most egregious nonsense to those who were educated in the Paris paradigm. Despine's writing suggests to me that he was incessantly trying to argue for the credibility of ideas that were a priori invalid to many of his contemporaries, and relentlessly insisting that if they were open-minded enough to observe the phenomena that he was studying, his ideas would be recognized and validated. Time after time he extols the remarkable phenomena he has observed, and assures the reader that if he or she witnessed them with an open mind, he or she would be convinced by the evidence of his or her own senses. He fills his monograph quoting correspondence that eulogizes him and his methods. However, his opponents appear to have regarded the phenomena he valued as "fruit of the poison tree," invalid because they stemmed from a priori improper observations, interpretations, and interventions.

Thus, Despine often sounds fairly desperate in his efforts to be convincing, and at times seems condescending to those who cannot accept his ideas. He sounds beleaguered, as if he is taking his best and last shot in an effort to make his mark. Despine is hoping to be worthy of his father, convincing to his son, and to win the respect of his community of colleagues. Although he succeeded in convincing many of the worth of much of his work, there is no evidence that he succeeded with many of those who mattered the most to him.

The contemporary reader will have difficulty understanding Despine's efforts to be scientific. Modern approaches to the question, "What is scientific?" will discard at once the majority of Despine's efforts to demonstrate what to him were profound truths. Even though Despine starts his monograph with, and in passing, makes many statements about scientific thinking and observation that are startlingly modern (e.g., he begins by outlining a proposal for a multisite study of both the healing action of spas and springs, to determine the similarities and differences of various institutions so that patients can be directed to the one whose waters best match their needs as well as targeted and parallel syptomalogical and interventional data collection on patients with idiopathic neuropathies), his scientific discourse is consistently flawed by his using information of different levels of demonstrability in order to prove his points, and by his failure to accord sufficient standing to the arguments of his opponents. That is, data from direct observation and the most abstract theoretical notions are combined in ways that fail to acknowledge that while the first can be authenticated by others, the latter remains hypothetical and cannot be regarded as firmly established (see Waelder, 1962).

Consistent with his mission and his awareness of the forces arrayed against his success, Despine's writing often has a painful and defensive tone. This grates against the sensibilities of the contemporary scientific reader, who is accustomed to a format beginning with more objective statement of issues being explored (with only passing mention of controversy and disagreement), and proceeding to a description of the methods and subjects used to do so, communication of the results, and then, in a discussion, raising various arguments pro and con his assumptions and his findings.

However, before decrying Despine for his defensiveness, it is important to appreciate that taking on anticipated objections in advance as one built one's argument was a not-uncommon feature of European scholarship well into the early twentieth century. Those who have studied Freud are well aware of how he endeavored to predict opponents' objections to his line of reasoning and to address them throughout many of his writings. Berman (1981) speaks of Freud's "defensive-polemic" tone. Although unusual by today's standards, Despine's tone was not that egregious for his era.

Also distressing is Despine's tendency to outline what would be the most appropriate and scientific stance to take, only to move away from the high standard he sets for his profession, without apparent awareness he has done so. His enthusiasm and driven determination undo him, leaving in their wake many frankly naive and erroneous statements. While it is tempting to draw analogies with Shakespeare's Polonius in *Hamlet*, who proposes a noble ideal of honesty, "This above all, to thine own self be true. . . ." while living a life of manifest hypocrisy, I do not believe that Despine was a hypocrite. His amazing dedication, his willingness to extend himself heroically for his patients, and his bravely holding to his opinions despite the controversies and attacks that assailed him, all suggest to me the unconscious self-deception of a sincere and basically honest individual who is so driven that he has lost considerable perspective.

In Despine's era, modern concepts of the patient role, boundaries, confidentiality, controlled studies, and transference and countertransference were largely unknown. Much of what was accepted as valid at that moment in time has been discredited, abandoned, or bypassed by subsequent scientific developments. I will try to illustrate these differences with a series of observations.

The patient role involves conforming to accepted conventions of disease, illness behavior and recovery. In any era, the prevailing models of the aforementioned concepts exercise a profound influence about how healers and the afflicted behave toward one another. For example, a half century before Despine wrote, the paradigm of possession was considered a valid explanatory concept in understanding why an individual suffered particular symptoms. Ironically, it was Mesmer's disproving the possession paradigm favored by the exorcist, Father Gassner, that played a major role in discrediting claims of possession as a legitimate way to express and explain one's distress. This paved the way for patients with similar symptoms to present those symptoms and their understanding of those symptoms in different ways (Ellenberger, 1970). Hence, within a matter of years, patients were described who were inhabited not by demons or ancestors, but by different supernatural entities and by secular ones as well. Exit demon and ancestor possession, enter multiple personality/dissociative identity disorder. The secularization or laicization of the Judeo-Christian possession syndrome brought with it a new set of paradigms, both early descriptions of dissociative disorders and the notion that problems with the quantity and distribution of a special magnetic fluid played a major role in health and sickness. Other analogies with electricity, the paradigmatic scientific advance of the era, suggested problems and remedies based on electricity-driven theories.

Despine was in fact quite modern for his era in working with all three of these new paradigms. He attempted an electric and magnetic treatment and he engaged the alters without moving to exorcise them. It is of note that Despine was an eclectic rather than a hard-core magnetizer. In fact, he had only encountered therapeutic magnetism a decade before he began to treat Estelle, when he was over 50. He saw himself as promoting a valuable synthesis of therapeutic modalities. However, with these theories and models came additional expectations of how afflicted patients might or should behave. Among them were the notions that somnambulists, high hypnotizable dissociatives, could both diagnose and recommend treatment for themselves and others, and that transposition of the senses was regularly observed. Hence, Estelle through her treatment with magnetism (i.e., a precursor of medical hypnosis) might predict what would be necessary to cure herself, offering Despine and the family guidance for her recovery. Despine only began to suspect a magnetic pathology when he learned that Estelle heard and appeared to interact with angels. Hence, Despine is eager to document that Estelle can see with her eyes closed, the visual sense having been transposed to her fingers. This was part of the clinical syndrome he believed Estelle to suffer. Vision was often said to have been transposed to the fingers or to the epigastrium. As ludicrous as this sounds today, it represented a significant stream of scholarly thought in Despine's era.

When patients come to understand that a particular set of symptoms is associated with a disorder, some feel, either consciously or unconsciously, a need to develop them. Cues, not only from general knowledge or other sources, ranging from most egregious suggestions to the most subtle of hints within a therapeutic or experimental situation, may imply to a patient or subject that he or she is supposed to do or say certain things that will be more expected or more favored by the healer or the experimenter. Thus, some of the phenomena that are manifested may not be intrinsic to the patient's disorder, but emerge as artifacts of the expectations that are conveyed, sometimes without any conscious intent to do so. The most notorious example of this phenomenon was the efforts of Charcot's residents to coach his patients on how to act in conformity with Charcot's theories. Once coached, without Charcot's knowledge, the subjects demonstrated findings consistent with Charcot's theories and models (Ellenberger, 1970). Babinski's discovery of this ruse discredited Charcot's ideas, and is often referred to as a landmark in the scientific study of mental illness.

Despine's patients, then, came to him with their own spontaneous symptoms, symptoms influenced by other care-givers and other sources of influence and information, and admixtures of the two. Estelle appears to have developed some phenomena in response to her contacts with other patients and in response to certain interventions and expectations. Also, it is not unthinkable that some of her symptoms indeed were manufactured by her to exert some degree of control over others. This complexity is the case with many patients in treatment, then and now, but because the major symptoms of Despine's era were often so dramatic and florid by today's standards, it may be hard to appreciate that the processes that gave rise to them are universal concomitants of both illness and the therapeutic or experimental encounter.

In Despine's era, the relationships between doctors and patients, unless there was great disparity in their social classes, were far more relaxed and social than would be considered ethical today. Patients might reside with a doctor's family for a period of time, and were attended to when they were symptomatic, not necessarily at a particular time of day. The medical practice of the era involved physical treatments for mental illness. There was no prohibition against physical contact per se, unless the nature of that contact was unethical by the standards of the time. Confidentiality as we know it today was nonexistent. Medical demonstrations at universities and various societies were attended by interested lay people, and some saw such occasions as educational or as entertainment.

Modern concepts of transference and countertransference had yet to evolve. The absence of such awareness could take forms that may seem grotesque to the modern reader. For example, Despine remarks that he allowed two patients with interesting symptoms to remain untreated for two years, to better study their phenomena for science. In contrast, Despine seems prepared to drop everything to attend to Estelle and her family, and dedicated considerable time to correspondence with her family members. It would seem that his countertransferential detachment with the first two cases is unacceptable, and that his work with Estelle and her family speaks of countertransferential overinvolvement. What both excesses have in common is that Despine consciously sees himself as doing something appropriate and worthwhile. In both instances his efforts to advance scientific knowledge provide a ready rationalization for his questionable stances. Dr. Fine's companion article in this volume addresses many of these issues as well.

Despite the impediments of inadequate theories, severe and entrenched psychopathology, and limited modalities of treatment, Despine was able to encounter and work with Estelle in a way that led to her stable and sustained recovery. How can we understand this outcome? It is helpful to appreciate that several hundred schools of psychotherapeutic thought and practice have existed, each with its proponents and grateful patients who have recovered under its aegis. Despine's diagnostic acumen was important—he was the first to understand the nature of her afflictions. Despine's dedication to Estelle within his model of treatment provided her with much needed encouragement and support. He allowed many aspects of her mind, expressed as personalities, to be heard out respectfully and brought into the treatment. He broke the previous pattern of relatedness between Estelle and the women in her family. Previously, she was a weak and dependent victim. Despine's model venerated the powers of the somnambulist, and accorded the somnambulist respect and credibility. Estelle found a voice for her assertiveness and mastery; she was able to achieve increasing individuation and self-efficacy, leaving childhood behind and beginning to work on developmental issues associated with adolescence. While at times Estelle and Estelle's alters could be tyrants, their mobilization and work enhanced Estelle's strength and competence.

A modern dissociative disorders expert or traumatologist might wonder if Estelle's traumata had actually been addressed, and point to subtle hints that some additional forms of trauma may have taken place (Fine, 1992; Goodwin and Fine, 1993). However, the trauma of losing her father and witnessing severe illness in other family members, including a sister, was more than sufficient to initiate a posttraumatic dissociative process. I wonder whether many of Estelle's symptoms mimic or were reactions to the illness behaviors that she witnessed in her own family. It is well known that children often believe others' misfortunes are the magical result of their own conflicted urges. I wonder as well whether Estelle found some way to blame the illnesses in her family on herself, necessitating her inflicting distressing symptoms upon herself in retribution, to assuage her guilt. For example, if she believed that she in some way was responsible for her father's death, Despine's withstanding her problematic feelings and behaviors without having been injured or driven away may have been a potent corrective emotional experience.

Psychotherapy research often demonstrates that while it may be difficult to demonstrate that one method of treatment is superior to another, the experience of the therapist and the length and intensity of the therapy may be crucial determinants. In Despine's work with Estelle, a very experienced physician worked arduously and intensively over many months with his young patient. Although one may argue that Estelle should not have been cured by a treatment many of the elements of which lacked scientific credibility, such reasoning discounts many very important and well-documented observations. Although theories of cure often are put forward in a manner that suggests that they must reverse the elements of a theory of etiology for the ailment in question, such claims are quite deceptive. The same disorder may respond to different approaches in different populations, suggesting that whatever is involved in the process of cure, there are many ways to initiate the helpful changes that restore a patient to health. For example, post-traumatic stress disorder (PTSD) might respond to modern trauma therapy in North American and European cohorts, but many Southeast Asian subgroups do not accept the basic tenets of this treatment approach, and may not be helped by it. In fact, they may deteriorate (Kluft, Bloom, and Kinzie, 2000). A Navaho might be cured by traditional ceremonies, chants, and sand paintings, but most individuals from most other cultures will not receive commensurate benefits from these culture-specific interventions.

It is often said that the efforts of professional healers are an art rather than a science. Part of the art is to provide the common features of all effective therapies; that is, to make available to the suffering individual a healing ceremony or a series of healing ceremonies performed by a healer legitimized in that society, a ceremony or ceremony that is governed by a belief system endorsed both by the healer and the patient. In the course of this healing ritual explanations are provided and emotion is expressed (Frank, 1973; Frank and Frank, 2004).

Despine's treatment of Estelle, however lacking many of its elements may be in currently accepted science, is rich in its provision of the elements of a successful psychotherapeutic experience. In a caring atmosphere Despine provided a treatment that embodied those elements incessantly day and night over a prolonged period of time.

I speculate that not only was the intensity of this treatment governed by several of the motivating factors noted above but there was also a special warmth and caring that suffused the endeavor. In this speculation, I suggest that in Despine, Estelle found a proxy for the father she lost so tragically in her youth and yearned to join in heaven. In Estelle, Despine found a child (and a family) willing to embrace him and his ideas with affection and enthusiasm, perhaps offering him some consolation for his scientific estrangement from his own son and whatever family issues had emerged in the wake of that situation and his chronic prioritization of his patients over his family (Fine, this volume). It appears that Estelle, entering adolescence, was more prepared to take leave of Despine than Despine was ready to forfeit his connection with Estelle. He tried to encourage her to return, and to participate in a contest to demonstrate the transposition of her capacity to read to another part of her body. In all likelihood some aspects of Estelle's personal revisiting of issues of attachment and of separation and individuation as she began her adolescent passage soothed her separation from Despine, while Despine was left to mourn the loss of an important relationship fraught with many levels of deep and poignant meaning.

When all is said and done, and all of Despine's personal issues and invalidated ideas are put aside, he stands revealed as a masterful healer, a largely unrecognized pioneer in the treatment of dissociative identity disorder and allied dissociative states and phenomena. He deserves our deepest respect. In another century and a half, when future colleagues review our contemporary theories and efforts, and put aside our personal issues and invalidated ideas, we will be fortunate indeed if we look as good.

NOTE

Richard P. Kluft, M.D., is clinical professor of Psychiatry at Temple University School of Medicine and on the faculty of the Philadelphia Center for Psychoanalysis. He practices psychiatry and psychoanalysis in Bala Cynwyd, PA. The author of more than 200 publications on hypnosis, dissociation, and the dissociative disorders and recipient of numerous awards for his work, he has been president of the International Society for the Study of Dissociation, the American Society of Clinical Hypnosis, and the Society for Clinical and Experimental Hypnosis.

WORKS CITED

Allison, R.B. (1974). A new treatment approach for multiple personalities. *American Journal of Clinical Hypnosis*, 17, 15–32.

Berman, E. (1981). Multiple personality: Psychoanalytic perspectives. *International Journal of Psychoanalysis*, 62, 283–300.

Caul, D. (1984). Outcomes. In Kluft (1984).

Ellenberger, H. (1970). *The discovery of the unconscious.* New York: Basic Books.

Fine, C.G. (1992). Estelle: a case of childhood multiple personality disorder in the 19th century. In J.M. Goodwin (Chair), Symposium on historical antecedents of childhood trauma, American Psychiatric Association Meeting: San Francisco, CA.

Frank, J. (1973). *Persuasion and healing.* Baltimore, MD: Johns Hopkins University Press.

Frank, Jerome and Julia Frank (2004). Therapeutic components shared by all psychotherapies. In A. Freeman, M. Mahoney, P. Devito, and D. Martin (Eds.), *Cognition and psychotherapy*, 2nd edition (45–78). New York: Springer.

Goodwin Jean M. and Catherine G. Fine (1993). Mary Reynolds and Estelle: Somatic Symptoms and unacknowledged trauma. In J.M. Goodwin (Ed.), *Rediscovering childhood trauma: Historical casebook and clinical applications* (119–131). Washington, DC: American Psychiatric Press.

Horvath, A.O. and L. Luborsky (1993). The role of the therapeutic alliance in psychotherapy. *Journal of Consulting and Clinical Psychology*, 61, 561–573.

Kluft, R.P. (1982). Varieties of hypnotic interventions in the treatment of multiple personality. *American Journal of Clinical Hypnosis*, 24, 230–240.

——— (1984). Treatment of multiple personality disorder. *Psychiatric Clinics of North America*, 7, 9–29.

——— (1986). Personality unification in multiple personality disorder: A follow-up study. In B.G. Braun (Ed.), *Treatment of multiple personality disorder* (79–60). Washington, DC: American Psychiatric Press.

——— (1991). Multiple personality disorder. In A. Tasman and S.M. Goldfinger (Eds.), *American Psychiatric Press annual review of Psychiatry* (Vol. 10, 161–188). Washington, DC: American Psychiatric Press.

——— (1993). The treatment of multiple personality disorder patients: An overview of discoveries, successes, and failures. *Dissociation*, 6, 87–101.

——— (2006). Dealing with alters: A pragmatic clinical perspective. *Psychiatric Clinics of North America*, 29, 281–304.

Kluft, R.P., S.L. Bloom, and J.D. Kinzie (2000). Treating traumatized patients and victims of violence. In C. Bell (Ed.), *Psychiatric aspects of violence: Issues in prevention and treatment* (79–102). San Francisco, CA: Jossey-Bass.

Kuhn, T.S. (1970). *The structure of scientific revolutions*, 2nd edition. Chicago: University of Chicago Press.

Luborsky, L., R. Rosenthal L. Diguer, T. Andrusyna, J. Berman, J. Levitt, D. Seligman, and E. Krause (2002). The dodo bird verdict is alive and well—mostly. *Clinical Psychology: Science and Practice*, 9, 2–12.

Schreiber, F.R. (1973). *Sybil*. New York: Henry Regnery.

Waelder, R. (1962). Psychoanalysis, scientific method, and philosophy. *Journal of the American Psychoanalytic Association*, 10, 617–637.

A Study of the Uses of Animal Magnetism and Mineral Waters in the Treatment of Disorders of the Nervous System Followed by a Case of a Highly Unusual Cure of Neuropathy

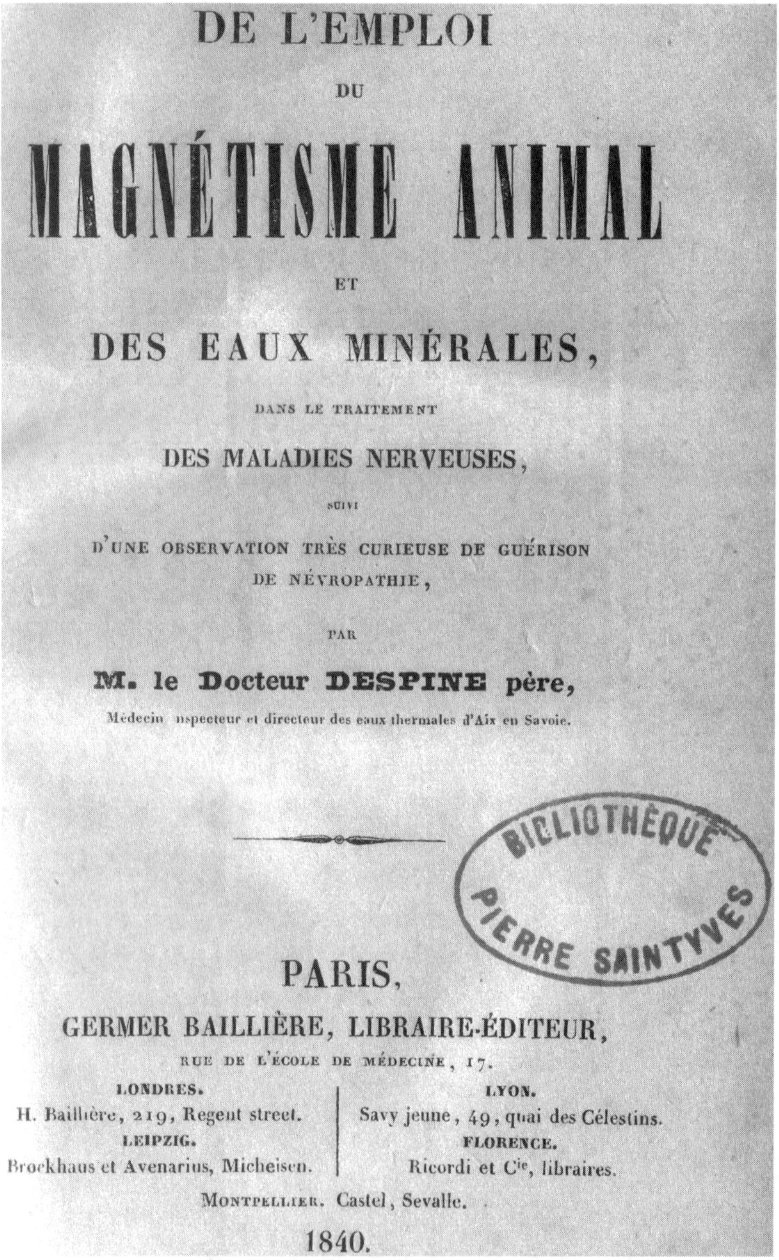

DE L'EMPLOI

DU

MAGNÉTISME ANIMAL

ET

DES EAUX MINÉRALES,

DANS LE TRAITEMENT

DES MALADIES NERVEUSES,

SUIVI

D'UNE OBSERVATION TRÈS CURIEUSE DE GUÉRISON
DE NÉVROPATHIE,

PAR

M. le Docteur DESPINE père,

Médecin inspecteur et directeur des eaux thermales d'Aix en Savoie.

PARIS,

GERMER BAILLIÈRE, LIBRAIRE-ÉDITEUR,

RUE DE L'ÉCOLE DE MÉDECINE, 17.

LONDRES.
H. Baillière, 219, Regent street.

LEIPZIG.
Brockhaus et Avenarius, Micheisen.

LYON.
Savy jeune, 49, quai des Célestins.

FLORENCE.
Ricordi et Cie, libraires.

MONTPELLIER. Castel, Sevalle.

1840.

Photo 3 Reproduction of original title page.

Source: From the personal library of Carole Koepke Brown.

Preface and Objectives of the Author

Much has been written on thermal springs, particularly in the last century. Some authors publicize their springs to potential customers, while others write to advance science. Although there are a fairly large number of general treatises on mineral waters and a few good monographs, we still need a good theoretical, practical work on medical hydrotherapy that says all that needs to be said, and no more. In addition to showing young practitioners the general rules of healing through the springs, this work should also describe the special attributes with which Nature has endowed certain waters. Finally, the study should indicate the ways that medical science has, through hard work, taken full advantage of the benefits inherent in the springs.

Nourished in the principles of the two most famous schools of the age,[1] I have taken from them a spirit of tolerance and impartiality. This attitude is rare in doctors today who, exposed to only one school of thought, have adopted its principles to the virtual exclusion of all others. The current zeitgeist is not one of observation—though the production of medical theories appearing over the past 200 years appears to demonstrate Hippocratic observation. These doctrines flourished when they were written, yet fell out of popularity soon afterward. Many writers seek the glory of publishing, but erudition today is rare. Many apparently new discoveries are, in fact, ancient findings. Moreover, our young writers—unfamiliar with the works of ancient and recent masters of medicine—often plagiarize unwittingly.

As for thermal springs, plagiarism will not be possible in a few years, since the local medical inspectors of each establishment must send, at the end of each season, an economic and medical report to the government. The government will send reports to the *Comité des Eaux Minérales de l'Académie Royale de Médecine* after the annual visits by the *Inspecteurs Généraux*[2] of the springs. Consequently, physicians serious about learning and healing the sick will have for their use a large collection of important documents.

Most unfortunately for science and humanity, the process of assessing and disseminating this very useful material exists only in France. Publishing my observations from Aix-en-Savoie[3] or reporting a lifetime of discoveries and conclusions to young colleagues will not

Photo 4 Portrait of Joseph Despine (1737–1830), father of Antoine Despine.
Source: From the private archives of Philippe Despine, Dijon.

fill this void. This monograph will be of some interest, nevertheless, for it includes some new and heretofore unpublished facts. These observations will give my young colleagues the opportunity to collect data for themselves and build on my work.

I have been involved with the springs in Aix for half a century, first under the tutelage and direction of the best of fathers[4] who witnessed the founding of this establishment and was its first *Médecin-Inspecteur*. Later, I visited the principal bathing establishments of Europe and assembled excellent notes.[5] For sites I could not visit, learned people collected materials for me. In many cases, I quickly concluded that physicians chose specific springs for their patients without knowing the establishment, often "judging the book by its cover." Their predetermined ideas were based on the climate, the name of the thermal sources, the age and frequency of use, or the soil from which the spring flows. Some physicians assume that unknown springs resemble springs they know. The two major mineral sources for Aix have long been the Sulfur-Spring and the Alum-Spring. They have brought us many patients. But we lose many patients because physicians who have not visited them know them for activation and excitation. Meanwhile, both springs work healing miracles daily either in achieving exactly what the treating physician intended or in accomplishing the objective in a completely different, unanticipated way. Why? Because our springs, like springs everywhere, have common properties that make them a universal panacea—especially when physicians on the premises tailor the springs to a particular need. In this way, springs can improve one's health. Unfounded claims, however, could cause a patient harm.

Among the abundant thermal springs in Sardinia, France, Switzerland, Germany, and Italy, many have a well-deserved reputation for combating certain ailments. Hence, the springs in Cachat, like those in Contrexeville, are specifically recommended for bladder and urinary tract ailments. The springs of St. Gervais in Faucigny, Laperrière in Tarentaise, Balaruc in France, and many magnesium springs in England are indicated for obese persons at risk for strokes and abdominal congestion. Finally, some springs saturated with carbon dioxide are advised to treat the upper gastrointestinal tract, since the springs effectively stimulate the appetite and gastric functioning: Bonneval in Tarentaise, Cour-Mayeur, St. Didier in Val-d'Aoste, Vichy, Bourbonnais, Auvergne, and Nivernais. These springs are also recommended for cardiac sphincter and gout. To improve the thermal establishment in Aix, we have imported all that seemed good from baths we visited. While the establishment has been in my care for more than 20 years, I also gradually introduced improvements based on my studies and clinical work as well as improvements inspired by famous physicians and scholars who visited us. Even the sick—understanding their own afflictions far better than anyone else—have frequently suggested therapeutic modifications that facilitated healing.

Thus we acted with constancy and relentless perseverance, making our baths a model establishment where almost every available treatment is offered. An excellent system organizes and centralizes materials that are convenient for and beneficial to the consumer. Recently Aix has become a European rendezvous where many of the ill who had sought healing elsewhere converge annually.

Over the past 20 years, some innovations I introduced (they were unknown or unused ancillary therapeutic methods) brought bitter reproaches from those who understood neither their importance nor my rationale. It is as if they find fault with anything they themselves have not done. Some persons said to me, "But, Doctor, our springs cured people very well in the past, without your hot-cold showers, electricity, and magnetism. They cured before you spoiled our inimitable God-given springs warmed by Nature. Our springs effected many marvelous cures before you introduced cold waters for mitigated showers, mild and temperate baths, and swimming pools. Our

Photo 5 View of Aix and the surrounding area in the early nineteenth century.
Source: Courtesy *Archives des Thermes nationaux d'Aix-les-Bains*.

baths, formerly taken at home, were undoubtedly as good as those you have created near the establishment."

That may be, I reply, but how many bathers came during the entire bathing season in the past?[6] Did people come to Aix for diverse illnesses? Certainly not. Furthermore, many more patients are cured now. All my colleagues bear witness to this. Also, I predict, without fear of being challenged, that should the least of these innovations against which you strongly protest be eliminated, our baths will begin to decline. Correspondingly, the whole town of Aix—the baths constitute the only industrial resource—will suffer.

Formerly, treatments at the springs in Aix consisted of baths taken at home, showers administered near the grotto of the Sulfur-Spring, cupping glasses or cornets,[g] and purgatives.[g] Thermal medicine featured such treatments at the time of Cabias (1622).[g] Until the Restoration in 1816, bathers were limited to bathing at home (using waters from either of the two springs), showering with waters from only the Sulfur-Spring, and, if the shower did not induce perspiration, submerging themselves in the Bouillon.[g] There were no steam rooms, no vertical or hard impact showers, no bells to alert the bathers, no adapters to localize the shower, and no temperate swimming pools in the large, main building of the baths.

Since the renaissance of our establishment, many services have been available: baths and showers at the Alum- and Sulfur-Springs; brushing,[g] massaging, and sweating rooms; and large pools for swimming and water gymnastics. A patient can stay for hours without becoming bored. There are masses of waters at all temperatures and impact waterfalls ranging from only several centimeters above the patient to a vertical force of 9 meters. We electrify, cauterize,[g] and, when necessary, undertake the most difficult surgical procedures. We administer baths with a sudden transition from hot to cold and from cold to hot that makes the person's entire constitution experience a kind of violent drenching that gives the patient vigor and strength and causes an overall physical upheaval. This shock changes the patient's habits and sometimes restores health immediately.

We still administer combination showers that gradually shift from hot to warm and to cold and then gradually reverse the shift. The showers can be regulated so that the water feels hot on one part of the body while feeling warm or cold on other parts of the body. Some patients are prone to strokes or aneurysms. Without the judicious use of ice, caution would have ruled out showers for such a patient. We use ice on the head while the rest of the body is treated with the action of our strongest, hottest water columns; in this way, we restore the person's health in these serious, difficult cases. Moreover, we heal greater numbers of sick now and successfully treat profoundly diverse ailments of patients who, in the past, would never have been sent to our baths.

The primary objective of this publication is to document remarkable cures done at Aix with and by the springs. A second objective is to benefit other thermal establishments by explaining our new methods to treat certain illnesses successfully, ones that had been considered to be incurable.[7] We have been led to these treatments by our systematic studies and tests, by wide-ranging considerations, by fortuitous circumstances, and by frequent, happy coincidence. Finally, this work should also benefit all humanity, promote my adoptive homeland,[8] and strengthen the reputation of an establishment whose beginning my father witnessed and whose growth and beautification I have seen. The thermal establishment has been—and will always be—the object of my partiality and concern.

Introduction

The preface gave the rational for undertaking this work and its objectives. In light of my age, I am not confident I will finish the project. With God's help, I will publish what is most important about the medicinal merits of our springs.

The chronic illnesses we commonly see annually at the baths in Aix can be classified into eight categories. I rank order them from the most predominant ailments to least predominant. I have always grouped them in my reports to the government[1] as follows:

- Rheumatism
- Skin ailments
- Lymphatic and strumous[g] ailments
- Ailments of the bones and joints
- Syphilis
- Paralysis
- Neuralgia
- Nervous illnesses marked by overall physical weakness, resulting from the absence, shortage, or maldistribution of nervous fluid.[2]

That being said, it seems logical to begin with the rheumatics, then take up the second group, and so on—to be useful to the most patients. Instead, I selected the noteworthy rather than the obvious. I begin with a case in the last category of disorders: nervous illnesses. Recently, the extraordinary cure of a young person from Neuchâtel demonstrates all of the most remarkable that medical science offers. Even if others consider the rest of my work to be useless and claim that it is like some political newspapers—not worth the paper it's written on—I am determined to publish this case.

I call my little heroine Estelle because her family desires to remain anonymous, fearing that too much publicity would compromise this child's future. This respectable, highly philanthropic family, filled with deep gratitude toward Divine Providence, will never hesitate to pay tribute to the truth whenever such witness is useful for science and humanity. I swear before God that the curious events I recount are absolutely truthful. The correspondence I maintained with this admirable family verifies this. The young patient's Mother authorized me to publish what I deemed useful. I selected, therefore, only what was necessary for a complete study.

Animal magnetism only accidentally became a part of my story. Today most scientific academies debate the subject—some treat it with disdain whereas others take it seriously. Everywhere, however, events related to animal magnetism occur, events noteworthy in the story of mankind. Members of learned societies express different opinions about these events that first startle and astonish (like sparks of light), but these events eventually cause greater confusion. Nevertheless, as these sparks multiply, they will one day enlighten the medical horizon. The Truth of these events will eventually emerge in all its glory to convince dissidents of their error.

Truth regarding phenomena related to animal magnetism seemed dormant in France until both new data and a prize proposed by Dr. Burdin of Paris[3] incited a polemic. The gauntlet, thus thrown down, was taken up by several zealous, learned physicians from the provinces. The reaction of the medical establishment to their interest must not have encouraged the physicians much, even though some other learned groups were more positive. The participants were still not deterred when the academy announced the donor would withdraw his money if no one had fulfilled the program's requirements within two years.[4]

If this competition were merely for curiosity's sake or were without consequence for science and humanity, I would eagerly encourage my young, worthy colleagues to withdraw from the competition and return to business back home, far from the confusion of the capital with its inflamed social and political passions.

This struggle, however, concerns a humanitarian cause—not idle passions. It concerns accepting or rejecting these somnambulistic phenomena as genuinely real, even though they are, indeed, truly rare and very unusual. Their rarity, uniqueness, and apparent divergence from the normative order of known laws of physics make them no less real. I was inspired to publish, without delay, one of the most curious cases ever recorded in the annals of springs. The timing of these findings in my clinical practice coincides exactly with the polemic about magnetism, as if to challenge or modify accepted theories about what happens near and within the nervous system during the normal state of health as well as during a disease process. Excellent treatises on anatomy and physiology have been available over the last few years. Today's scientists have focused on the encephalon[g] and its annexes, though researchers of the last century had already worked with them. Even since Thomas Willis (1621–1675) many have zealously and successfully worked on numerous experiments that gave excellent results. But have all authors proceeded correctly? The best course of action—the only valid one in these investigations—is observing, observing well, and observing often. Furthermore, conclusions should not be drawn nor theories be derived until the premises are absolutely indisputable.

The admirable works of Gall[g] deserve consideration as a great contribution to science; yet the structure of an organ does not explain everything. A veritable Tower of Babel developed when mechanistic explanations based solely on knowledge of anatomy and organ function were used to explain the phenomena of life. People no longer understood one another.

Scientists—making up a first group who subscribed to this mechanistic perspective—assumed that a human being had some kind of internal electric battery that initiated and sustained physiological processes. In that way, these particular scientists claimed to explain everything.

A second group of scientists attributed acts of intelligence, physical movement, and organic and physiological processes (whether instinctive or not) to a spiritual impulse the soul gives to all (human) machines. This soul, an expression of the Self in each person, is an immaterial reality emanating from God. It seems to manifest itself primarily in the higher functions of intelligence, while secondarily invigorating the biological automaton to which it is only temporarily attached, as if only passing through.

Finally, a third group of scientists recognized in human beings organized matter, an intelligent soul, and a "vital instinctive force." Organized matter gives humans their form whereas the intelligent soul presides over functions of understanding and animates the whole machine. The vital instinctive force, given by Nature to all creatures to assure their preservation, is born in them and dies with them. Moreover, this instinctive force is, strictly speaking, both the vehicle of animal life within the individual and the force that drives all interrelated actions.

This concept of instinctive force represents a mid-ground explanation between the first and second groups of scientists. (The first group is materialistic and mechanistic; the second group believes in the impact of an immaterial soul.) The vital instinctive force would constitute what some philosophers have called the "soul of the animals." This vital force or "glue" would be the essence of each human and would be at the intersection between body and soul. At first, body and soul seem so incompatible in light of their nature and function. Yet the material and immaterial must be intimately linked to form "living man" and endow him with reason. This principle of instinctive vital force, set in motion by yet ungraspable circumstances, may activate our physiologies. This force, that may synchronize the functioning among all our parts, is responsible for the organizing life force in all living beings.

My book is not a philosophical treatise. Unable to treat such questions, I entrust such work to metaphysicians and theologians. I do not take up these delicate, profound matters—the most difficult of human sciences. In the treatment of nervous illnesses, however, I have observed some most extraordinary events related to this instinctive force. It is difficult for me to accept the existence of these events without recognizing an intermediary element between the spiritual soul (the emanation of the celestial author of all creation) and the body (the material whole and the soul's home on earth). This hypothetical intermediary element seems much more probable than the spiritual and material components working alone. With this intermediary instinctive force, we can explain (or at least make sense of) an endless number of vital phenomena. I limit my role in this study, though, to that of historian, as I do not explain the phenomena. I must warn my readers that I propose nothing in this work that I have not witnessed (I took all precautions against fraud and trickery). In fact, in every country where such patients were observed, their phenomena are more similar in character and intellectual expression than they are dissimilar. These shared patient behaviors alone should establish the validity of the phenomena. Moreover, if patients were deliberately faking these phenomena, they could not be sustained so consistently everywhere.

With concerted effort, I recorded the patients' verbatim verbal expressions in this extraordinary magnetized state—the only means to capture with precision what these patients feel and experience. In other words, their words and tone genuinely echo their intimate feelings. Their vocabulary is quite remarkable and their style is concise—the patients use no superfluous words. Their self-presentation is like a mirror that shows them as they really are, without disguise or makeup.

To see what I have seen, others must be as involved as I have been. They will obtain the same results, since the essential character in all these patients is the same, modified only by differences of temperament, education, and social position. Finally, analysis of these patients' expressions and thoughts shows traces of wisdom and insight not found in patients with other disorders. Study of such phenomena may answer, at least in part, some essential psychological questions asked by ancient and contemporary philosophers.

Scientists have been unable to agree on a theory explaining how nerve impulses transmitted by the brain, the cerebellum, and their annexes sustain life in a healthy person. It is not surprising, then, that for an unhealthy person we are unable to agree about nervous phenomena—so variable and disparate. Even though we have much scientific and historical data, neither abundant data nor derivative experiments advance our knowledge of the phenomena! Although one can attribute certain mental illnesses to an identified neuropathy of the central nervous system, many other mental illnesses go misunderstood. In these cases, autopsies fail to corroborate the most skilled practitioners' predictions. We could conclude, then, that most cerebral ailments (mental illness) are not caused by actual brain lesions.

My narrative about Estelle demonstrates that some pathological cases apparently suggest the largest, deepest of brain lesions—without there being any lesion whatsoever. Estelle's general paralysis lasted for several years, confounding all resources, means, and medical prognoses. Her paralysis would cease, come back an instant later, disappear again, and reappear soon afterward. This unpredictable behavior was, though, impacted by the application of someone's fingers, including Estelle's, on certain regions of her body—specifically nerve trunks, nerve branches, or nerve ganglions.[5] The little patient had no rational reason to suspect that these spots were significant; her instinct alone inspired her to touch them. A paralysis of this sort will, no doubt, find no plausible explanation in the medical theories filling the halls of academe. The *Tò Theyon* of the Greeks explains it no better, and modern doctrines (founded on chemistry, mechanics, and inflammation) are of no use. Vitalism alone permits us to approach the question. Objections and difficulties remain, however.

For a rational explanation about the phenomena I discuss, one must make three assumptions. First, the brain secretes a very subtle fluid that eludes our mechanical measuring devices. Second, this fluid leaves the brain and is disseminated throughout the whole body. A third assumption is that large quantities of this fluid are lost at salient points on the body—just as electric current moves to the protruding points of a conductor. Next, all fluid that does not remain at those points would return from the periphery to the system's center, the common sensorium. Here, the surplus fluid would mix together with newly secreted nervous fluid. Until death, therefore, a perpetual circulation and feedback loop continues.[6]

The great discoveries of the Galvanis and Rolandos, of Ampère, Magendi, and Flourens support this data. But how does this circulation work? Does it happen only in one unique way? Does it simply follow the nervous system and its innumerable nerves? Or does the nerve fluid, moving through the main conduits of neurocommunication along large neural tracts and their numerous branches, unexpectedly jump or move sideways? If so, our present knowledge of anatomy cannot explain such action. The nerve fluid may behave like the electric current in the atmosphere, attracted to the tip of a lightning rod, then usually following a conduction chain to a common reservoir. But, the current also often goes left or right, zigzagging and impacting on the parts nearby that have similar characteristics.

The facts reported here, the clear, precise explanations somnambules give on their inner experiences, and future studies of this curious state may answer this perplexing question about nerve fluid. Careful, sound study of magnetism will convince skeptics who will agree that magnetism elicits artificial somnambulism, whose character and physiological phenomena are identical to those of spontaneous somnambulism.[7] Magnetism will provide frequent opportunities to observe, verify, and study many features of the somnambulistic state. The features have seemed extraordinary only because those who are able to appreciate their full value have not seriously examined them.

Before I studied magnetism as a physiological phenomenon or as something that causes striking pathological or morbid symptoms,[8] I had frequently witnessed reactions of the nervous system that were inexplicable (according to what was taught in medical school). I observed these inexplicable reactions dispassionately and objectively. I studied them, reflected on them, and found I could not reject them. The more I examined them both in isolation and in context, the less I understood them. Nevertheless, the data suggested that the phenomena must have a natural cause. Therefore, I concluded that an event, always occurring in the same form (however unusual it first appears) and occurring with some regularity, must be governed by natural laws that are fixed, immutable, and real.

Every year during a relatively short season, I see male and female patients of different ages and beliefs from all over the world. These patients suffer from various ailments typically treated at the springs in an empirical rather than a rational manner. Early in my practice, I was asked to see a fairly large number of invalids in pain, suffering from nervous ailments. The doctors attending these patients diagnosed almost all of these ailments as rheumatism. These problems intensified when treated with showers and hot baths. True rheumatic pain, however, were it muscular or articular (all things being equal), would heal quickly with this treatment. Consequently, I concentrated my studies and research on the discrepancy between the diagnosis and the symptoms. Studying nervous disorders to identify their genesis and different treatments led to the remarkable results I develop below.

Illnesses of the nervous system have always been the despair of medicine. These ailments exhaust doctors of high repute who enjoy a large clientele. Their demanding responsibilities leave them little time to devote to those afflicted by nervous disorders, even if these respected doctors are well motivated. These ailments also frustrate doctors with a limited clientele because the time necessary for treatment inadequately compensates the doctors. Finally, nervous afflictions drain the zeal and curiosity of new, young doctors. Indeed, without clinical experience, beginning doctors know these disorders only through book knowledge and a few hospital cases. New, unusual phenomena continually frustrate these inexperienced professionals. They do not yet understand that treating nervous ailments—more than treating any other sicknesses—demands patience, perseverance, and time.[9] This rare, good fortune to treat many patients with nervous illnesses allowed me to study the illnesses in their different forms and follow their progression as well as the impact of our treatments. Moreover, I have benefited from curing several patients who had been considered incurable.

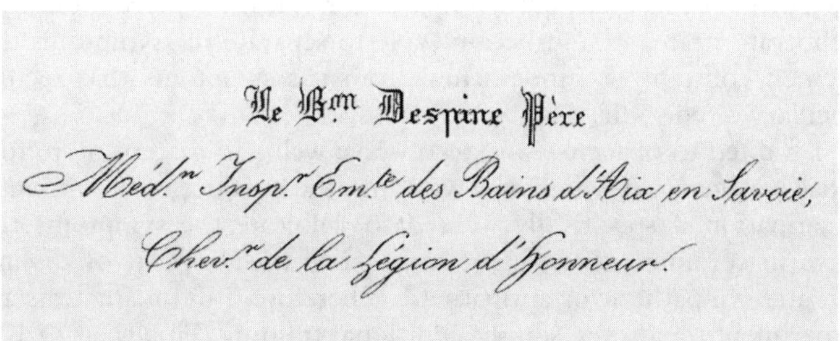

Photo 6 Antoine Despine's professional calling card.

Source: From the private archives of Philippe Despine, Dijon.

Observing cases of catalepsy and somnambulism gave me the opportunity to compare what I observed with what others had seen and described. First, these events interested me greatly—as they do others—for scientific reasons. My observations suggested that the evolution of the body's nervous phenomena is analogous to the progress of electrical phenomena. Although I did not find a perfect correspondence, I was convinced there was a significant relationship between them. Hence, I did tests, and I succeeded, even where others had failed! The success was enough to motivate me. I sharpened my focus and enthusiasm for what I believed to be my own discovery. From the outset, therefore, I avidly seized each opportunity to shed light on doubts and uncover new information.

I particularly focused on the inner workings of the body and on which circumstances naturally facilitated the phenomenon of transposition of the senses.[g] I sought to understand why this transposition would occur in one part of the body rather than another. Moreover, I explored whatever might explain certain rare, unusual phenomena in my patients such as localization, intensification, and numbing of sensation. These phenomena seemed new and interested me more than catalepsy itself.

A fortuitous combination of circumstances advanced my work. In quick succession, I saw five patients stricken with more or less advanced stages of catalepsy (an illness authors criticize and describe so diversely). Quickly understanding the patients' immediate problems and indigence,[10] I warmly welcomed them into my home for continual, repeated observation. I could, then, personally verify their extraordinary comments about their condition. I studied them to know and analyze their presentation and the cyclical nature of their attacks. I also intended to help these patients by having them study each other. By facilitating their interaction, I learned more about their deliberate strategies and became able to separate the symptoms that seemed constant or fundamental from the symptoms that seemed ancillary or episodic.

I wanted to observe—and to observe well—to discern more fully what was genuine from what was only the product of deception or imagination. I specifically wanted to delineate the symptoms that were in common, that were constant or varied, similar or dissimilar, and sympathetic or antipathetic. Therefore, I did many tests and experiments, and even sought additional training. Finally, as early as 1820, I included a series of curious events belonging to this unusual order of pathophysiological phenomena in my annual governmental reports on the medical season of the springs in Aix. My perseverance in conducting this kind of study had led me to fairly significant data on the nature, presentation, and treatment of these phenomena. The very encouraging results of my work are (1) the simplification of the treatment of illnesses of the nervous system, notably those of diverse spasmodic neuralgias; (2) the increased awareness by critics who had considered these illnesses marvels, artifice, or sorcery; and finally (3) the increased awareness by critics who had considered these ailments incurable and believed either current prejudice or outdated misconceptions.

Thereafter, I continued to gather any information—including similar, although not identical symptoms—that might further clarify the subject. For a long time I intended to publish my observations but without opportunity. I spoke of these curious phenomena with several doctors I knew in Paris, Lyon, Geneva, and a few other large cities. Several of these men observed or cared for patients with the same kinds of illnesses. Like me, these physicians were surprised by the unusual nature of these ailments and were struck that public opinion viewed these apparently miraculous events with disfavor. They encouraged and applauded my effort. Not one, however, believed he had enough data or enough observations to publish his own work.

In the meantime, I had given all my notes to the knowledgeable, unassuming Alexandre Bertrand,[11] who honored me with his confidence and friendship. This worthy colleague (unfortunately for science, he died young!) wanted to include my notes in his extensive comparative study of catalepsy, ecstasy,[12] magnetism, and the varying stages and types of somnambulism he treated *ex professo.* This would have been an encyclopedic resource for these phenomena. The premature death of this esteemed friend, however, left this work unfinished and probably lost to science.[13]

I then offered these same materials to other doctors involved in magnetism, but they did not respond. Since I am convinced that what happened to the great work of Bertrand could happen to my own observations on the laws that seem to govern the unusual phenomena of catalepsy and other illnesses of the nervous system, I decided to publicize the case study of my young patient from Neuchâtel and my discoveries about her illness and treatment. Phenomena of this nature can be described or explained sufficiently well only by those who have actually done the experiments. I am likely to experience many a difficulty, criticism, and setback by publishing these improbable events. These events are so important for medicine, however, that I feel obliged to publish them, albeit reluctantly. How I present the phenomena here, though, demands the readers' indulgence.

This account involves a child stricken with almost general paralysis for several years. She allowed herself to be cauterized by fire, irons, and the most painful caustics. Although these treatments prolonged her suffering, she never complained. She did all this because her loving and concerned mother convinced her that a cure would result from these very difficult trials. Of all my cases, I believe that her case best proves the honesty and integrity of my work.

It involves a child adored by a mother who never left her side. This sweet and likeable young lady, born and raised in Paris, had been suffering to varying degrees for between eight and ten years. Because of her age and youthful frankness, she cannot be suspected of emotionality, conceit, or the wish to cause a sensation. Indeed, her extreme innocence made her incapable of deliberate deception. Her family, who had been faced with misfortune since 1830, had better things to do with its money than to waste it in Aix. The thought of being in debt to their family was already too painful for both mother and child. That reason alone guarded against fraud or any inappropriate behavior. Moreover, trial by fire is not a process to which children or even adults submit without definite need.

Before beginning Estelle's story, I present my reasons that these disorders are little known and that their phenomena are generally viewed with prejudice and disfavor. Fortunately for humanity, real cases of catalepsy, as described in scientific accounts, are relatively rare. Moreover, special competence is necessary to treat these patients effectively. An independent, motivated mindset allows one to find the required time, seize all opportunities to make keen observations, and remove all obstacles. These qualities are rare, however, even in physicians, unless their concern for science and humanity is a priority as they work, study, and research. Not surprisingly, doctors are called in only for extreme cases. Illnesses of the nervous system typically occur acutely at irregular intervals; their most remarkable, fleeting phenomena happen in the middle of the night. Most colleagues who sent me these patients had not witnessed their crises. Often their accounts were assertions made by others. Finally, I have identified what some magnetizers called epilepsies as only simple convulsive ailments (or perhaps the nervous disorders we treat at our baths). Although magnetism cured these illnesses, few were real epilepsies.

For other reasons as well, physicians are unaware of the causes or types of nervous ailments that weaken a patient. Spasms, loss of consciousness, tetanic stiffness, loss of sensitivity, lethargy (and so forth) are seen equally in cases of catalepsy, hysteria,[14] and epilepsy. Even learned men in the field often confuse these illnesses by their overlapping symptoms, considering them varieties of epilepsy. No one is comfortable speaking openly outside the home about these troubles. Families wait patiently for time and age to take care of the problem, calling a physician only when the illness has become chronic. In the past, those convinced the problem was unsolvable hid the sick in monasteries until they died and then spoke no more of them. In other cases, the patients were hidden at home in a domestic prison they left only when they died or when, no longer wanted, they were sent to a faraway hospice. A heavy pension was less onerous to an eminent family than to have a physically or mentally disabled person perpetually at home. Oh, how many men and women (but especially women), victims of these pitiful ailments of the nervous system and made incurable by such steps, might have been cured. Another reason people know little about these unusual ailments is extreme mobility of the nerves[15] that makes patients impatient, irascible, and very difficult to control. Finally, there is complete ignorance about the laws governing nervous fluid and its distribution (most physiologists even doubted its existence for a long time). All these circumstances undermined the successful treatment of these ailments many considered obscure and doubtful. By default, the resulting therapies were merely symptomatic, and physicians had little taste for them.

Meanwhile, some isolated, remarkable cases of nervous ailments appeared. The seriousness, organization, and extent of care depended on the physician's knowledge of and interest in the patient. Often, the patient became a subject of experiments and tests. Some patients seemed so extraordinary that an inexperienced physician, struck by the unusual phenomena before him, could do nothing but observe and study them—if only out of curiosity. Yet an unbiased, independent, passionate, and selfless physician with the necessary leisure to study the problem with great care was rare. Sometimes, however, an extraordinary man appears.

Pétetin was one of these. Unfortunately for science, this illustrious writer was unable to finish his work and continue his unusual research because of circumstances in Lyon at the time of the siege.[16] Nevertheless, his thoughtful study in physics led to his understanding the therapeutic effect of electricity on such nervous indispositions. Moreover, he devised a treatment (literally founded a doctrine) that healed nervous illnesses with the mysterious electric fluid. My many studies have not yet refuted Pétetin's immortal work. His data on animal electricity will long serve physicians seeking to understand more thoroughly—and, more promptly, effectively treat—these curious illnesses.

His cases, however extraordinary they may appear, are no less true and no less possible today. Without exception, I have seen them all in more than 20 similar cases in my practice. Moreover, however rigorous or unusual the treatment Pétetin described may seem, it still works best—even according to the patients. Distinguished men of modern medicine such as J. Frank, Dumas, Lordat,[17] and other illustrious professors from the best schools recorded that they witnessed phenomena similar to Pétetin's. These men published their works for the world to see. Although their motives were unclear, these same scholars and celebrated professors spoke disparagingly and ironically of Pétetin. In fact, he was treated as a dreamer and madman. Even his integrity and character were suspect. Time, however, corrected these senseless allegations and doubts. For a long time, I experienced much the same criticism (see my endnote 8). Recently, however, the authentic, complete, and unexpected healing of Mademoiselle Estelle and several other similar patients at Aix (under my direction and that of my estimable colleagues), have rehabilitated my reputation. These patients may also comfort others who seek relief from similar symptoms after ordinary medicine fails them.

We find observations of nervous disorders throughout human history, in all countries and languages. Germany has countless reports. England, despite its skepticism, is beginning to acknowledge similar events (see my endnote 15). Now Dr. Elliotson in London is actively promoting his research on animal magnetism and the phenomena of somnambulism. Many ecstatics have been reported in Voghera, Caldara, and Bologne. If one puts aside all religious or mystical influence, the physical pain of these ecstatics is similar to that of ecstatics I treated in Aix.

The phenomena of magnetism will enjoy the same renown and rehabilitation when medicine accepts magnetism as a preferred therapeutic method to treat nervous illnesses and when physiologists classify them. The symptoms that emerge under magnetic influence are no more marvelous than the spontaneous phenomena observed in catalepsy and certain other nervous states. All of these curious symptoms must be considered as physiological phenomena. If they vary, this variation is not due to fundamental differences but to age, temperament, or habit. In the treatment, these nuances are of little importance. This explains the extreme impressionability to magnetism of people stricken with nervous ailments: sleepwalkers, cataleptics, epileptics, hysterics, and hypochondriacs.

Magnetism today differs somewhat in theory and in practice from what it was for Mesmer. If the doctrine differs, however, the phenomena remain the same and therefore magnetism, in its essence, remains the same. Mesmer associated his discovery with equipment that soon proved to be unnecessary. Perhaps he considered these devices indispensable, but if he intentionally claimed the equipment was necessary only to sell it, he would be a charlatan. *L'Académie Royale des Sciences et de la Faculté de Médecine de Paris* challenged his work in a committee review based much more on partiality and party spirit than on devotion to science. Afterward, the study of mesmerism was advanced and perfected. Later, its application was simplified by the works and courageous research of de Puységur, Deleuse, Bertrand, Dupotet, and Foissac. Consequently, magnetism is beginning to reappear with some standing on the world stage. Moreover, it has recently been supported by important scientific discoveries in galvanism,[18] electricity and the brain, and by knowledge of the influence of electromagnetic currents[19] on the production of muscular movements and sensitivity. The study of magnetism will make rapid progress. The gains would be greater and faster if we could use the machines and instruments of physics to investigate the nervous fluid (the vital fluid itself). Judging by progress made in electricity, physiology, and mechanics (all within less than a century), we may soon understand this fluid much better.

There has been a rebirth of important studies since the general pacification of Europe—discoveries of Aldini, Galvani, Amoretti, and Pétetin on animal electricity, and those of Ampère, Prévost, and Dumas on the irritability of muscular fiber subject to electrical currents. We also have the research of Gall, Rolando, Spurzheim, Flourens, Magendie, and others on the brain, its dependencies and functions, and that of Avogadro and Michelotti on the action and galvanic capacity of metals. Manni and the modern physiologists have worked on cases of asphyxia and apparent death.[20] Further investigations by these researchers are shining a new light on science.

Let us observe, therefore, and examine without any preconceived ideas. It is impossible that so many well-educated men, strangers living far from each other, could see such marvelous, yet identical phenomena without some truth in what they observe. Could this similarity be insignificant? Likewise, it is impossible that patients who are strangers to one another could have agreed to trick physicians everywhere for the simple pleasure of mystifying them. It is unlikely that these sick patients would suffer—subjecting themselves to cauterization and needles—without reason. Therefore, let's see what these marvels are all about and judge for ourselves.

Because of ideas such as these, the *Académie Royale de Médecine de Paris* thought it best to withdraw its rejection of Mesmer's discovery and thus finally judged animal magnetism worthy of new research as more data strengthened the case for its phenomena. The 1831 reports of the *Académie Royale*, June 21 and June 28, declared that there was actually something to magnetism: it was not, as some had believed (or pretended to believe), fraud or a figment of the imagination. They finally established positive doctrines that, henceforth, serve as a basis or a point of departure for new research (see P. Foissac, 1833, *Rapports et discussions*, p. 199). Magnetism has been debated so vigorously that a sensible person can no longer reject the obvious. Now that this great step has been taken toward the science of magnetism, it will not be difficult to climb to the top of the ladder.

As for me, dear reader, I have read many works on the nervous system, magnetism, mental illnesses, and the various aberrations of human understanding. Moreover, I applied accepted principles to the cases I observed. I gathered considerable data and research and admit, albeit reluctantly, that some reasoning and deductions are questionable! Therefore, the most reliable reference is still Nature. Her truths are accessible to whoever wants to study her. She is like a reliable resource whose tenets are tested and valid. Nature does not reason according to fallacious theories but presents facts. This manner of reasoning is always irrefutable. It involves simply observing events well in the light of day, with both care and discernment.

The great mistake of authors who teach something new is to dogmatize too much. To present original material easily, they believe they must reduce science to a few principles (the fewest possible), as if to consolidate it. Unfortunately, these scientists draw conclusions before having enough data to support them. Fear of undermining old beliefs or of destroying principles about long-standing academic physiological knowledge has often caused researchers to abandon their experiments. But will Nature behave similarly? Her laws are simple while diverse across individuals, types, and races. How could someone ever meet two completely identical patients since, in the immensity of the forests, it is difficult (impossible, perhaps) to find two identical leaves? If individual variation within its classification is accepted as true for the simplest bodies of Nature, all the more reason to recognize individual variation as true in illnesses, including those of the nervous system. The form, manner of being, and phenomena of these disorders are very mobile and as variable as is the great imponderable nature of the fluid that is their material cause. Hence, it takes a discriminating mind to examine and compare often enough to identify what is distinctive in nervous disorders and to differentiate what is pathognomonic from what is genuine. In this way, we participate in the evolving understanding of these illnesses.

The many nervous ailments observed in my practice enabled me to differentiate between the data that are constant and the data that point to differences. I drew conclusions from the unchanging data. They led me to formulate certain principles or laws that apparently govern the nervous phenomena of natural somnambulism. The same laws also apply in cases of ecstasy, catalepsy, hysteria, and other nervous disorders.

The diverse nervous phenomena produced by magnetism resemble convulsive ailments, but the magnetic phenomena occur by accident. The convulsive ailments are found in practically the same circumstances as the magnetic phenomena and appear to be subject to the same laws. Hence, I must make this parallel known, along with their overlapping similarities. I have only begun. I leave the task of completing it to men of science after me.

A Curious Case of Neuropathy with almost Complete Paralysis Using the Springs, Electricity, and Magnetism

In medicine, as in history, there are those events that astonish and seem to be outside the natural order of things. Such is the cure I describe here. I ask the reader's indulgence for the lengthy details. The medical case requires them, science demands them, and those who did not witness the cure will need the details to find these events believable. Neither a physiologist nor a medical practitioner would believe in such a cure if only its astounding points were highlighted rather than its details given chronologically. Therefore, only by the progressive, sequential unfolding of the phenomena does one have incontestable proof that this story is neither fairy tale nor fiction.

Medical History

Mademoiselle Estelle L'Hardy[1] was just a few months past her eleventh birthday when she arrived in Aix on July 15, 1836. She was referred to me by Dr. De Castella in Neuchâtel, one of the most knowledgeable Swiss physicians and chief of staff at *l'Hôpital-Pourtalès.* In a letter dated July 6, 1836, this esteemed colleague wrote me, "I am sending you an intriguing young lady who, following a softening of the spinal cord, was stricken with a treatment-resistant paralysis. She is seeking your medical care and healing waters. Her treatment will most likely be quite lengthy and demand many special accommodations. You will probably have to apply new moxas[g] or cuppings[g] along her spine. The last application had some impact, and I would have repeated it were it not to delay what we hope to be her fruitful trip to Aix. If she recovers, your springs will never have brought about a more wonderful cure. It will be a true medical miracle, and I will enjoy publishing it, adding the history of the illness from its beginning.[2] I have no doubt that you will derive great satisfaction as you follow this particularly remarkable case. The extreme sensitivity of the skin over her entire body and the pain caused by the slightest pressure on the ridges of the vertebrae will strike you."

This young lady, gifted with a rare intelligence and a kindly character, is a member of one of the most distinguished families in the Evangelical Pastorate.[3] From her early years, care was taken with her education. By age four, Estelle was already reading fluently many kinds of works, and she was a respectful and dutiful child and pupil.

At age five in the midst of one of the deadliest measles epidemics in Paris, she contracted the disease; she developed brain fever that included elevated temperatures and other frightening symptoms. Treated by Dr. Capuron, she recovered completely except that she remained subject to very frequent headaches. When the 1832 cholera epidemic broke out, she was still living in Paris. It killed her father whom she cherished. It violently struck her Mother and her younger sister Blanche. Estelle herself, generally in good health, though frail and delicate, did not get sick but she was intensely distressed by the death of her father. Her sensitive and loving nature was overwhelmed by these events. Emotionally, she became highly impressionable, and physically she became hypersensitive. It was thought that living in the elevated areas of Switzerland, exercise in the country, and recreation would be the best antidote for rheumatic migratory pain, chronic rheumatism, torticollis,[g] and daily headaches that were becoming increasingly intolerable. Unfortunately, these measures had very little impact. Good days alternated with bad ones. Since the child was eating, drinking, and sleeping, however, the situation was going reasonably well.

Then, on November 27, 1834, Estelle was involved in an apparently insignificant accident that developed serious consequences. Estelle was now a few months past nine years.[4] She was playing with a girlfriend her age. They were holding each other's hands, pulling backward, tugging and jerking each other when Estelle's friend suddenly and unexpectedly released both of Estelle's hands. Estelle fell flat on her bottom. Even though she fell only a short distance—from a standing position—the jolt was such that the shock was felt in her entire small body and particularly strained the dorsal and lumbar spine.

At the time, this incident alarmed no one. Estelle, not wanting to get her friend into trouble, did not talk about her pain. Her family, not wanting to encourage self-pity, essentially ignored the incident and rarely talked about it. Her condition concerned them only when

Estelle began losing strength, complaining of pain in her back and between and above her shoulder blades. She was looking less and less well; her stomach was swollen and taut. Toward the middle of December when Estelle developed chills and a fever, her family began to take her health quite seriously.

In addition, she was urinating infrequently. Dr. De Castella, who was called in, detected hematuria.[g] Estelle also experienced an overall bloating and some disturbance in her digestive functions. The patient was prescribed mild, temperate baths with very good results and generally remained happy and vivacious. Consequently, her minor discomforts were treated as childish and insignificant: she was reluctant to walk, to stand up, to stand still; she refused to have her pulse taken and to have the doctor examine and touch areas that bothered her—her back and stomach. She also claimed to feel symptoms that we could not understand or make sense of. For all these reasons she was called "The Spoiled Child from Neuchâtel."

Her overall condition, meanwhile, deteriorated. Her headaches intensified; she developed a cough[5] resembling the barking of a dog, developed a sensation of suffocating, as well as acute pain in her stomach, bowels, and chest. She also reported extreme sensitivity all over her skin, along with other unusual, puzzling symptoms that would take too long to enumerate. It was thought that forcing her to walk in her room and to get fresh air would improve her condition. The opposite actually occurred. After the first attempt, there was such an intensification of headaches and breathing problems that no one dared to force her. The convulsive cough and breathing problems returned predictably each day, lasting from between 4:00 and 5:00 in the afternoon until between 10:00 and 11:00 at night.

Several physicians did face-to-face examinations whereas other physicians in Paris were consulted long-distance. Because the physicians could not diagnose this illness with its serious symptoms, they resorted to prescribing symptomatic interventions. As a consequence, Estelle was alternately treated with rub-downs, medicinal poultices, leeching, soothing medicinal syrup, vesicatories,[g] bitters, and mustard plasters, to name a few, in a crisis-oriented rather than tactical manner. Among the many therapeutic modalities, only vesicatories[g] and sodium chloride taken orally appeared to be effective. Meanwhile, Estelle's tongue, which until then had been somewhat coarse, grainy, and scabby, even blackish at times, cleared up. Around April 5, 1835 a convalescence and gradual improvement of sorts began and continued well into May. When the child could walk a bit, her complete healing seemed possible. This improvement, however, was misleading; toward the last days of May it completely stopped. No one knew why or how. Soon thereafter, she regressed, and her previous symptoms returned and worsened. Through the end of June 1835, then, a healthy diet was initiated along with additional vesicatories,[g] arnica,[g] and magistral[g] syrup. There was some improvement.

In July 1835, Madame L'Hardy took Estelle to the baths in Baden, Argovia, where she herself had gone for treatment. Once there, Estelle took only those baths usually used for personal hygiene. Each time she used the baths she felt feverish, felt chills down her spine, and experienced increasing weakness in her legs; consequently, the baths did her no good. Her headaches became increasingly intolerable. Meanwhile, though, the peacefulness at Baden, the warmth of the season, the recreation, and the change of routine, seemed to benefit Estelle. When they returned to Neuchâtel, it was considered wise to resume the vesicatories[g] and bitters. She often took donkey rides. For a time, she was comforted. On the anniversary of her fall,[6] however, Estelle had cramps. She then rejected all food except a little bread and milk. Nervous, persistent fits of coughing soon followed. At that point, the child stopped walking on her own. Initially, she maneuvered by holding onto furniture she could reach. Soon thereafter, she needed crutches. Finally, she could not walk at all, despite new vesicatories,[g] prolonged balsamic massages, and fortifying, vigorous foot baths.

The winter passed as well as could be expected. At the end of January 1836, however, the convulsive cough intensified to such an extent that the patient suffered horribly. Witnesses feared she might actually suffocate. At this point Estelle asked Dr. De Castella to allow her to eat a little snow for which she felt an overpowering craving. He agreed to her request, almost in desperation, monitoring her so that she ate sensibly. Imagine the surprise of all involved when—after two days on this diet—the persistent and alarming cough completely disappeared!

Meanwhile, the pitiful little invalid found herself forced to stay in bed, despite her natural vivacity and feistiness, because her legs were weak. In March 1836, her paralysis worsened, and her spine became so painful that when Estelle sat on her bed, her body would bend over in an arch. It seemed to collapse on itself. While examining her spine, Dr. De Castella thought he recognized the beginning of gibbosity;[g] thus, Pott's Disease[g] seemed imminent to him, if not already in its early stages.[7] He had recourse, therefore, to the only treatment means that until then had been truly effective for this illness: two stone-made cauteries[g] were opened on her back, one on each side of the apparent protrusion. The headaches ceased immediately. Her stools, which had been abnormal for a long time, remained whitish. Her back pains, which returned every night at 8:00 or 10:00, seemed to increase markedly. Her relief, thus, remained incomplete. It was decided to try a moxa[g] whose slower and deeper action might more effectively draw out toxins.

Thus, on April 5, 1836, two moxas[g] with a diameter equal in size to a 100 centimes coin were placed near the base of her spine. The child endured the painful application (it lasted 12 minutes) with resignation and courage beyond her years. The desired result did not occur. From that moment on, our curious patient became completely immobilized in her bed; it was no longer possible to move her without causing her horrible suffering. In the beginning of June, however, De Castella paralleled the impact of the actual cautery by applying 12 cuppings[g] in a single day. There was not even the slightest relief for the invalid, and it seriously unsettled those involved with her.

The caregivers considered repeating the moxas[g] and introducing the *Eaux Thermales d'Aix* since the therapeutic methods used until then (cauteries, various baths, massages, diverse oral potions) had produced few advantages. Everything about treating this sickly patient and extraordinary disorder seemed confusing and abnormal. Since the season of the springs was well underway, it was prudent to take advantage of the baths immediately, yet reserve the option to apply the moxas[g] again if the waters did not facilitate Estelle's healing.

Estelle and her entourage left for Aix and, after five full days of travel from Neuchâtel, arrived on July 15. During the trip the patient was transported in a large, custom-made willow basket with a flat bottom since it was impossible for Estelle to sit upright. Even though she lay flat, mattresses had to be stuffed all around her to guard against her being tossed or shaken.

So she was tucked in her basket, with eiderdown quilts on all sides, despite the excessive mid-summer heat. Moreover, the windows of the carriage had to be firmly shut during the entire trip. At each stop, the straw stretcher-like basket was taken down from the carriage and carried to an inn. No one except Estelle's Mother or her maternal Aunt Julie Lardy[8] could touch the invalid without making her scream. Every time they stopped, the transfer of this peculiar basket became a novel attraction, drawing a crowd of inquisitive spectators. Generally, these loiterers had no sympathy for the young invalid and her entourage who would have gladly done without these scenes.

PRESENTATION AND ASSESSMENT

When they arrived in Aix on July 15, I was quickly called to this unusual patient's side. I carefully examined Estelle. I did not despair, even after I had taken her history, and I recognized in her an illness most serious and obscure in its nature. I was heartened somewhat by the emotional strength of the child and by her intelligence—neither had weakened despite her loss of physical strength. But I could not ignore the danger signs she presented at the time of my first assessment.

Physical State

Estelle had to spend the entire day in an absolutely horizontal position in bed. She was so sensitive to temperature changes that she was wrapped in cotton padding and eiderdown, despite the intense summer heat. She was placed on a bed of wool and feathers, warmed with pitchers of hot water. On her own, Estelle could not hold up her head; it had to be continuously supported and padded with cushions of feather and horsehair. She was lifted from her bed only to have it made. Her Mother alone could move her without causing her terrible suffering. In order to reduce her pain and distress, several days would often pass when no one dared to move her at all.

When her neck was not too stiff, Estelle managed some lateral movements of her head. She could not be seated in an armchair or a couch at more than a 35–40° angle because her head would fall back from its own weight—her neck and back had no strength. She then would let out piercing screams. It was, if you will, like trying to hold up a cherry by its stem.

The body of our poor little invalid, laid out immobile in her bed, remained in whatever position she had been left. Her face, like the rest of her skin, was chalk-white even though, from time to time, her cheeks colored slightly. The patient coughed intermittently, always in fits; moreover, the cough was dry and nonexpectorant. Her pulse could not be taken without making her anxious and irritating her. She did not sweat profusely, and her hips were always as cold as marble despite all measures taken to maintain warmth there.

Estelle ate nothing until noon; she could eat neither meat nor broth. Constipation was customary, and each day the invalid herself asked to use the clyster-pipe[g] (a procedure all children detest). Although she drank beer occasionally, she drank neither wine nor liqueur. Ice cream and sherbets, however, delighted her and never caused any problems.

Emotional State

Estelle's mental state was slightly better than her physical state. As a result of her long suffering, her emotional stamina was weakening. While she slept, dreams, fantastic visions, and hallucinations reminded Estelle of painful memories from the previous evening or from her past and led her to periods of terror, fright, and constant dread—particularly when alone. A mouse, spider, fly, butterfly became (in her hypnogogic[g] moments) robbers, monsters, or vampires. She believed they came to her at night with their terrible grimaces to devour or frighten her. The least expected noise caused her heart palpitations, anxiety, and unspeakable terror. At times, these unusual symptoms were accompanied by a distinctive emotional state (observed by Estelle's Mother) that Estelle herself could not make sense of. It was common for Estelle to have a conversation or hear something of great interest to her, yet minutes later be unable to recall the smallest detail! Similarly, when we carried her on walks, she was interested in all she saw around her and chatted about it; yet on her return Estelle often seemed amnesic for her experiences. Or, if some traces of the experience remained, they were furtive like an escaping dream.

It was believed that the illness was going to her brain and that it would soon end with the onset of dementia or stupor. No one mentioned this concern to the child, for fear of upsetting her. Estelle herself made a point of not speaking of it, despite her painful thoughts in this unusual state because she did not want to seem crazy or admit that she had no memory of many events. Her young pride would have been offended. Profoundly affected by her endless suffering, Estelle had come to envision life and death with equal indifference. Estelle lived only because her death would hurt her wonderful Mother.

Etiology and Prognosis

What were we to conclude from an illness as complex and inscrutable as this one? Obviously, this case was difficult. If certain findings indicated a grim prognosis, others did not and, in fact, seemed somewhat favorable. Hope, therefore, was not completely lost for these reasons and conditions:

- Estelle's healthy constitution;
- her deterioration began—quite remarkably—only after repeated emotional afflictions in her childhood;
- her fall (November 27, 1834) led to the overall weakening of her legs but without very apparent gibbosity[g] or protrusion on her back;

- the cauteries,[g] moxas,[g] and other techniques for her spine, even though they were applied near the assumed center of pain and were designed to draw out toxins, turned out to be almost completely useless. Without these established methods that had proven to be quite effective in Pott's Disease,[g] other lesser known but proven methods to promote healing from this disease would also be considered essentially useless;
- the striking contrast between (on the one hand) the excessive sensitivity of her skin over the spine and pelvic area and (on the other hand) the absolute insensitivity and immobilization of these areas to deeper touch, yet they remained without limpness, flaccidity, or noticeable emaciation of her muscles;
- the sharp contrast between Estelle's intellectual faculties and her physical state;
- Estelle's overall reactivity to cold, a highly unusual response in compression paralysis; the quite notable power of her will—quite remarkable in a child not yet 12 years;
- finally, the perseverant, tender and assiduous care of the best of Mothers, who did not once leave her child. Moreover, Madame L'Hardy was very determined not to interrupt the process at Aix (aware that all other medical interventions had failed), if her Estelle were benefiting. She was motivated by even the smallest hope of improvement, even if her expectation fell short of complete healing. Even a little progress might mean a bearable existence and a normal life span for her child.

Because I could not ascertain the etiology[g] of the disorder, I thought it wise to begin by treating Estelle symptomatically, as my other colleagues had, and then to act *à juvantibus et lœdentibus.* In the course of this treatment, a fortunate coincidence one could call providential led us to try animal magnetism as a therapeutic method.

TREATMENT

First Stage: July 15–December 20, 1836
Treatment and Diagnosis in Aix-en-Savoie
(Includes Endnote 1: Estelle's Six Demands)
Beginning of Treatment

Between July 15 and August 15, 1836, Estelle took about ten showers without their exhausting her. On August 15, she became nauseous, suffered a violent headache, and developed red blotches all over her body. On August 20, the very first perceptible spontaneous movements in her legs appeared. The patient was then able to lift herself a bit in her bed. On August 21, she could lift her head up and even hold herself upright in a sitting position, bracing with both hands. This improvement continued gradually and perceptibly until the first of September. At that time the weather turned rainy and cold and noticeably affected Estelle. One day, when she left the showers without taking enough precautions, a sensation of cold made the blotching disappear and her improvement stopped completely. She even regressed.

On September 15, we began electrotherapy[9] along with hydrotherapy. This combination of modalities quickly produced very good results. Her strength was restored, and her whole body progressively experienced a very perceptible improvement. Therefore, we continued hydrotherapy and electrotherapy together. Since her paralysis had set in, Estelle had needed to use her hands to move her legs. But on the first of October, the patient began to move her legs spontaneously in bed without the use of her hands. By November 1, she could get up on her knees. She held herself up on her legs while supporting her body against the bed, her arms firmly holding onto nearby furniture. Then cold weather—as severe as the previous year—set in prematurely. Because of this unfortunate change of climate, Estelle's progress halted.

Crucial Episode regarding Estelle's Neuropathic Disorder

An unusual incident, however, shed light on Estelle's uncommon affliction. Estelle, her Mother, and her Aunt Julie intended to return to Neuchâtel before the extreme cold of winter. They planned to spend the winter there and return to Aix in the spring to complete the treatment. (This plan was particularly desirable to Estelle's Aunt who had left her own Mother under someone else's care. Aunt Julie's desire to be with her Mother continually troubled her.) On the day scheduled for their departure, I was pointing out to Madame L'Hardy the considerable contrast between Estelle's emotional and intellectual strength on the one hand, and her physical strength on the other hand. Because of this sharp contrast, I had reason to be optimistic about Estelle's recovery. The same considerable discrepancy had characterized the illness of Monsieur de Vallenstrale, a young Swedish Baron sent to me by Baron Alibert the preceding year. Like Estelle, he was a paraplegic. I was remarking to Madame L'Hardy about his extraordinary success at Aix because of his courageous perseverance in our thermo-therapeutic methods when she suddenly remembered an incident she had omitted from her account of Estelle's history.

"Last April," she told me, "I was very surprised when my daughter, who had always been easily frightened and who had become even more so since her illness, asked me to leave her alone on a few occasions! When I asked her to explain such an unusual request she said, 'Momma, you wish it to be so! I am going to tell you, but please don't get angry. I am quite sick; but, you mustn't get too worried. For a while now, God, touched by my suffering and wanting to console and sustain me in this long trial, allows me to hear the singing of celestial spirits. Every evening, at the same hour, it is as if they seek me out to comfort me with their melodious chants. They are so beautiful, Momma, these hymns of the Angels, that nothing can describe them. I have never been able to understand a single word, yet they speak to the heart, and I am very sure that Papa sings with them!! The least noise interrupts this celestial concert; this is the reason I want to be alone—furthermore, if it weren't for you, my dear Mother, I would want to unite myself with them quite quickly—very quickly—and forever. Yes, Momma, your Estelle would like to unite herself with these celestial spirits whose canticles so charm her. Besides, her state of suffering is so painful that she doesn't know how much longer she can endure!!!' "[10]

Until now, Madame L'Hardy had been careful to say nothing about Estelle's episode to a physician, fearing he would believe Estelle to be delusional. Madame L'Hardy avoided talking about it with anyone, including her daughter, fearing that Estelle's premonition about her own death would come true. Meanwhile, this conversation increased my hope; Estelle had experienced phenomena that belonged to the pathological state nosologists call ecstasy. I had often observed it in my cataleptics. From that moment on I suspected more than ever that Estelle suffered from a functional ailment rather than an organic ailment. This functional ailment was related to catalepsy although presenting somewhat differently. Therefore, I worked at convincing Madame L'Hardy not to return to Neuchâtel. To ensure Estelle's cure, I insisted on the importance of staying in Aix. In addition, I stressed the benefit of summarizing any features of her ecstatic state and the importance of being alert to any new phenomena related to ecstasy and somnambulism.

Madame L'Hardy was struck by this new perspective on Estelle's illness and by the specificity of my description. Since this approach was completely new for her, I had to provide ample explanation. To acquaint my patient's family with the astonishing (although quite natural) phenomena that characterize these particular states of the nervous system, I spoke to them of catalepsy, somnambulism, and magnetizers. I even offered to let them witness all these phenomena, more or less nuanced, by introducing them to two ecstatics[11] whom I was treating in Aix at that same time. I had observed in these patients partial paralysis, heightened sensitivity in certain areas, transposition of the senses,[g] and countless other little known phenomena (often treated as fictions) that piqued the curiosity of these ladies. The connection between these stories and Estelle's condition was still not clear, but I thought something might be gained.

I spoke to them informally of the instinctive foresight somnambulists have about personal matters and of the immense asset this ability can be to a physician responsible for a perplexing, difficult case. By engaging this instinctive foresight, patients become the best guide to direct their treatment, since in this state they speak frankly about their feelings and bodily sensations. Although perhaps having doubts, these ladies were curious to see a demonstration of my amazing assertions. I rose to the challenge and soon showed them all I had explained. I then offered them the works of Pétetin, Bertrand, and Foissac to read. Soon, their confidence and trust in me and the process became so complete that Madame L'Hardy begged me to try some magnetic passes on Estelle. She was hoping that, if we could induce a state of somnambulism, Estelle's instinctive intelligence may take hold and facilitate self-preservation. Then, with her newfound lucidity—so extraordinary—Estelle might propose a treatment that would be quicker and more effective than those already tried.

Very outspoken, Estelle disagreed with her Mother: "I do not want to be magnetized," she said forcefully. She feared that she would prescribe for herself new types of moxas,[g] more intense electrification, and Scottish Showers,[12] the thought of which sent chills up her spine. She also feared becoming an automaton, acting without intelligence like a machine that blindly obeys the directives and will of others. Estelle also dreaded being overly reactive, perhaps walking and running without being able to stop and not knowing what she was doing; she had witnessed these behaviors in Henriette Bourgeat, one of the two ecstatics.

To no avail, I endlessly explained to Estelle that her present state—not unlike an amputee—was far from that of running without being able to stop. If the need arose, we could immediately stop our intervention. I repeatedly explained that, during induced or spontaneous somnambulism, one can confidently ask the sleepwalker about the course of treatment because the self-preservation instinct serves as the great regulator. This instinct is the prime directive for a somnambulist. My explanation did not reassure Estelle. Indeed, for a long while, I got nowhere with her. Finally, however, giving in to her Mother's pressure, she permitted me to try some magnetic passes—on the condition that I stray neither from the instructions she would give me in advance nor from the series of pre-planned questions Estelle or her Mother spelled out. Estelle insisted that we ask her only these six questions[13] while she was magnetized:

- Does the electricity suit you?
- Can you walk?
- Should we continue to brush[g] and massage?
- Shall we continue the fumigation?[g]
- Must you stay long in Aix?
- What must you eat and drink?

I conceded.[14] The first session was set for the next day.

Meanwhile, I forewarned Madame L'Hardy that, despite my conviction that Estelle was a suitable candidate for magnetism, I could not guarantee its success. I knew neither how long it would take to induce the sleepwalking state nor how long it would take to bring Estelle to the degree of lucidity that Madame L'Hardy wanted. I added that the art of healing with magnetism was still in its infancy and that I was neither a magnetic healer by profession nor a magnetizing physician. I was simply a medical observer who saw his patients without bias or emotionality. I told her that I admire the beautiful phenomena of Nature, studying and analyzing them perhaps more meticulously than many others. I also mentioned that I had applied my ability to acutely observe Nature in order to study animal magnetism for the benefit of my patients. This is no different from the way I normally record and systematize the causes and treatment of a disorder. In the same way, I had learned to work from an etiological or curative perspective. Finally, I told Madame L'Hardy that I was neither a fanatic about magnetism (as some in Aix claim) nor did I see miracles everywhere (as some disciples of magnetism swear blindly on their master's word). Instead, I told her I was an eclectic physician, selecting whatever could be useful for my patients, irrespective of its source.

I was able to convince this excellent Mother that somnambulism was still essentially and historically a mystery; indeed, most of its laws were still unknown. Moreover, one did not achieve somnambulism at will, nor would a somnambulist necessarily present in the same way each time. Finally, I made Madame L'Hardy understand that, since I was at the advanced age of 60 my nervous fluid had diminished. I could no longer depend on it for my magnetic passes to act vitally on Estelle.[15] Moreover, I could not guarantee that she would enter the somnambulistic state. Yet I could assure her, keeping in mind my long experience and Estelle's circumstances, that there would be no harm done to her if this little known method were tried. Mindful of these concerns and according to these principles, we began.

Second Stage

December 20–December 22, 1836
Treatment in Aix-en-Savoie

On December 20, 1836, the entire Aix Basin was covered with about 30 centimeters of snow.[16] For several days the temperature had hovered right below zero, but then it consistently fell each night so that after Christmas the temperature had rapidly dropped to −5° to −15° C.

Baseline State of the Patient

Estelle L'Hardy had suffered for a long time: 25 months ago, she fell; she had started her sixth month of treatment, and each day she improved somewhat. However, progress was so very slow that Estelle still had to spend the whole day in bed except for the 10 to 20 minutes we tried to sit her in a chair or on a small sofa. Moreover, it was necessary to prop her up with cushions and steady her with uncomfortably tight straps around her knees. We had to lay her on an inclined surface covered with cotton batting and then struggle to raise it to an angle of 50° to 60°. If the angle of the plane were any higher, a cushion had to be placed under the crook of her neck for support. She could hold this position for only four to five minutes.

Her Mother and the attendant usually responsible for showering her were unable to touch or move Estelle without causing her great suffering. The skin on her back, chest, and over her whole scalp was excruciatingly sensitive, as well as the skin on her arms and legs with their network of nerves and blood vessels. No porter in the thermal establishment had yet been able to transfer her from her bed to a chair and back again. Thus, I recommended that a little canvas stretcher and two frames be made for her to shower. One frame was horizontal; hinges on the other frame functioned to raise or lower the back support, according to Estelle's needs. While in the bath, the child had for a while been able to use a chair for paralytics by tilting it and by supporting her head with a small horsehair pillow (it was lodged in an indentation on the upper edge of the back support). At the same time, I ordered a little portable chair to be made for Estelle's walks; it was suited to her size and constructed according to the same principles in order to provide comfortable support for her head, back, and feet—all at once.

Despite her noteworthy improvement from treatment in Aix by December 20, 1836, it did not matter how one placed Estelle in bed in the evening: the next day she was found in the same position. During the 12- to 15-hour rest her state demanded, she was immobile like a block of marble. Moreover, as soon as winter arrived, we needed new eiderdown quilts. To stop her from shivering, we doubled the cotton and eiderdown in her bed. Her whole body was pale and sickly; her derma, thickened and puffy. Even in bed Estelle could still not move her legs without using her hands: as a matter of fact, she could barely even spread her toes or flex her feet. She did so only with considerable effort by twisting them inwardly. All this suggested that the musculature on the outside of her legs was deadened and flaccid. Moreover, all these spontaneous movements would stop or be reduced to half their usual number if Estelle tried to do them on demand.

Use of the Waters

Estelle's showers customarily lasted between 60 and 90 minutes. During the first third of the bath, the water level was gradually raised to her waist. During the second third she was massaged. During the last part the water temperature was raised from 28° or 29° C to the natural temperature of the waters (34°–36° C). This process continued until Estelle asked that it stop. However sensitive Estelle was to the cold, she could not tolerate a bath above 30° C for long, and she shivered in water below 28° C.

Use of Electricity

Electrotherapy was administered either in the morning at the time of her shower or in the evening before or after her walk. It usually lasted 20 minutes: ten for electrification by sparks and ten for electric shocks, also called discharges from the Leyden jar. Typically, we administer approximately one hundred shocks at a speed between two to six turns of a glass wheel from Paris, 76 centimeters in diameter and two rows thick.[17] Estelle endured these difficult procedures with courage and resignation. The spasms spread only from one foot to the other, and consequently only in the direction of the "vital circle of action of her legs."

Progress Made

Our young patient benefited physically from more than five months of electrotherapy, including three months of regular electrification. Even though we were still far from our objective, Estelle's morale was much improved. Estelle was now rarely depressed. She no longer wanted to die, and she was often cheerful again. She still dreamed almost every night, however, and at the slightest surprise she reacted emotionally and experienced heart palpitations. Insignificant mishaps startled her. At night, it took her hours to fall asleep because she feared snakes, beasts, and robbers. Her Mother, who usually slept at her side, had to chatter constantly until Estelle made the transition into sleep, overpowering her daughter's feelings of panic and terror. Once asleep Estelle was peaceful, but troubling dreams interrupted her rest almost every night.

Activities

While awake in bed, our invalid continually amused herself with reading, writing, drawing, or creating cutouts with pleasure and dexterity. Estelle reported not feeling too disheartened; she was becoming accustomed to her monotonous life.

Noteworthy Phenomena Exhibited by Estelle to Date

Estelle's treatment with showers and baths, although very hot, had not caused her to perspire abundantly, as would be expected in the showers at Aix. She experienced moments of light perspiration at infrequent intervals, always with some obvious benefit.

Lipothymy and Other Nervous Conditions

At times recently, when electrotherapy[18] was too strong for Estelle, it provoked a very distinctive state of choking followed by lipothymy.[g] Other times, the shocks (which acted only on her legs) nonetheless caused a reverberating sensation all the way to her throat. This experience exhausted her. At such a time, she begged us to stop quickly because, she said, while crying, "That bothers the nerves in my stomach." We laughed at her expression. Yet her words described a reaction, in her nervous system, to what was going on in two very different states. At this time, we simply believed she experienced a lipothymy. We had not yet realized that she was on the cusp of a developing ecstasy as well as catalepsy.

Diet

During the first part of the cure at Aix, her diet had been more or less the same as her diet in Neuchâtel. Estelle ate nothing until lunchtime: she had no appetite all morning long. She never ate meat or drank broth, for both completely disgusted her and actually "hurt" her. Her regular diet consisted of herbal soup, boiled eggs, and some garden vegetables. She had eaten a lot of asparagus in Neuchâtel, but in Aix the asparagus season was over. Each day she drank milk, coffee, and ate grapes, oranges, crumbs of bread, and ice cream. Every day she needed the clyster-pipe[g] without which no bowel movement would occur. Putting on underwear, a dress, or a camisole was like asking for the world. The smallest movement caused her unspeakable pain.

With hydrotherapy and electrotherapy already in her regimen, we came to the most unusual stage of the treatment. Magnetism would provide its therapeutic usefulness, and somnambulism would finally evolve, with all the marvels of its singular state.

Last Stage: December 22, 1836–End of April 1838 Estelle's Magnetic Cure in Three Phases

It would take too long to describe all the remarkable things that happened to our curious invalid as soon as magnetism became part of her treatment. Every day there were new wonders connected to one another like links on a long chain. It is essential to understand these connections to have a sound idea of the process of Nature in the gradual development of the phenomena of ecstasy, catalepsy, and somnambulism. Therefore, I group them into three phases in order to explain the evolution of the cure.

Phase One[19] covers the time Estelle felt the need to use ordinary magnetism, meaning that the magnetic state was induced by someone else. Phase Two covers the time spent in Aix when Estelle, able to do without a magnetizer, was able to magnetize herself spontaneously. Finally, Phase Three covers all that happened after her return to Neuchâtel until today.

Phase One of the Magnetic Cure in Aix-en-Savoie:
December 22, 1836–January 22, 1837
Includes Endnote 2: Mother's Transcription

December 22, First Session Pressured by Madame L'Hardy, Estelle first allowed me to try magnetic passes on her.[20]

The patient was stretched out full length on a small sofa except for her head that was raised on cushions at an angle of 20–25°. She was wrapped in her double eiderdown, her feet tucked into a bag of feathers; a stoneware jug filled with boiling water was placed against the bag. Without touching Estelle, I began with full longitudinal passes from her head to her feet, sometimes doing throws and stops. My hands remained five centimeters above her body.[21] These throws and stops were done at the top of her head, at her temples, eyes, cheeks, along the carotid veins in her neck, and on the pathway of the vagal nerve. Sometimes I stopped above the middle of her chest; other times I stopped at her hands, lightly squeezing her thumbs.

Before beginning the magnetic passes and even while I began them, Estelle laughed, chatted, and joked about this treatment with all those present (including me). I was quite surprised when, after about 20 minutes of magnetism, my young patient suddenly stopped laughing and joking, to remark, "Monsieur Despine, your magnetism will work much better than I first thought. I sense that your fluid has an effect on me that I never experienced before——[22] I'm beginning to see little bluish grains in front of my eyes——and when your fingers pass over my eyes the grains become all red. But when you do throws, the grains become like flashes——Keep going for a few more minutes——I sense that your fluid is putting me to sleep in a gradual and quite extraordinary manner——." After one and a half hours of uninterrupted magnetism, Estelle said to me, "That's enough for today——I'm going to wake myself up——I'll see you tomorrow——at the same time——please——."[23]

An instant later the patient woke up, experiencing a small shudder all over her body. She yawned, stretched her arms, and—completely amazed with herself—began to say to her Mother, "Ah!——Hello everyone!!!! But Momma, where am I? What happened?——I feel as though I've just come out of a deep sleep and a long dream——but, I feel well——very well——I feel very differently than I felt earlier. Oh! If it is magnetism that causes this shift, then I no longer fear it——see you tomorrow, Monsieur Despine, see you tomorrow, please——see you tomorrow——see you tomorrow."

December 23, Second Session The next day Estelle was brought to my home[24] for electrotherapy. Her impressionability had increased so much that I had to reduce the intensity of the electric discharges and administer them with only two turns of the glass wheel to avoid renewed spasms and lipothymia.[g] After leaving my home, she was taken for a walk in her small portable chair. At 8:00 in the evening, I went to her home. Like last night, we used magnetism, but now her response to the passes was much more rapid.

After 15 minutes of magnetic passes, the "small grains of fire" reappeared. Consequently, to hasten and deepen her sleep, Estelle requested full longitudinal passes from head to toe, my insufflation[g] on her fingers and palms, a little magnetic influence on the top of her head, and a lot on her face. Next, she asked that I do passes in this manner: rapidly moving my hands over her chest and epigastrium,[g] lingering over her abdomen and then prolonging the passes over her legs, and stopping at her knees and instep. "How warm your fluid is, Doctor," she would say to me, "it goes right to the head, overheats it first, and then the fire moves down the spine, on both sides, and spreads throughout the body.[25] If we proceed tomorrow like today, I think I will fall completely asleep for I am already three-quarters asleep now——[26] Ah! Monsieur Despine, your fluid is so hot! It penetrates to my bone marrow!!! The bottle of hot water is no longer necessary for me, nor the eiderdown——Take them away, please, until I have come out of my crisis[27]——Make no noise, please——You are whispering, Ladies! But, I hear you quite loudly, you know——Everyone is wearing me out here except Monsieur Despine. Please step back. If you wish to ask me something do so through Monsieur Despine——and I will respond through him——Hide the light, please; it is bothering me.[28] Tomorrow you will have to take it away much earlier and not make any sound at all——Momma!——Momma!——[29] Step back, please, you are hurting me.[30] Oh, my good Monsieur Despine!!! Please thank Henriette Bourgeat for letting herself be magnetized in my presence. I will be grateful to her for the rest of my life——."

Elaborating further about what she was feeling, she added, "Now, Monsieur Despine, we must continue the magnetism every day. At this point, it is more helpful to me than the baths and steam rooms. You must also continue with the electricity, but not too strong, please! Once I return to Neuchâtel, I will have to be given a lot of asparagus——But, my dear Monsieur Despine, I was tricked before when the asparagus was cooked in beef broth!!! Humph! You don't know all the harm that caused me! Promise me that I will not be tricked again——I ask for only what best suits me——In the future, give me all I ask for except fried meatballs. Since I like them a lot, I might ask for them; however, absolutely no one should give them to me——You will have to resist all my requests, no matter how earnest. Oh! This time I see that I'm moving into somnambulism and I think I will soon be completely cataleptic——wait, Monsieur Despine, it seems that I can see you even though my eyes are shut——and you, Momma, yes——oh! I see you there with your hands crossed and your arms close to your body——as for you, Mademoiselle Amélie, you are more difficult to see!!! More passes, Monsieur Despine, they are plunging me further into sleep——Take and hold my thumbs often.[31] I sense that your fluid has more effect that way: it moves up from the arms to the head, then moves down to the pit of the stomach, making it sort of quiver. Next, the fluid travels down to the base of the spine where it divides in two to go down to the feet——But let's not bother with the legs today; rather, let's try to bring on true sleep by continuing full passes that completely clear my head. It feels odd when you touch the chin and the neck! My eyes shut like a box——Meanwhile, I still see you, Monsieur Despine, and when your fingers pass over me, they seem to be on fire, like real lightning bolts."

At 10:15 Estelle woke up suddenly after a short silence. Her demeanor exhibited the same phenomena, the same disposition, the jolt, and the same surprise as the previous day.

December 24, Third Session Estelle had a very good night. The following morning she asked for a respite from electricity but not from magnetism.

I followed the same steps as the day before, mindful that I must observe her rules and instructions——. The "grains of fire" appeared at the sixth minute; after the eighth minute, she seemed to taste or feel some residue in her mouth; after 10 minutes, Estelle experienced light convulsive jolts in the arms and in the legs; after 15 minutes, she saw and made out my hands exactly (her eyes were closed). After 35 minutes, she accurately guessed (at that moment) what one person witnessing the session was thinking and reported it in a loud voice. After 40 minutes, she had a nightmarish vision. It represented a horrible figure that filled her with fright. After 50 minutes, she saw her Grandmother in her usual residence at Peseux, near Neuchâtel. After 60 minutes, she saw sparks coming from the tips of my fingers. After 15 additional minutes, Estelle had a new vision that filled her with joy and hope. In this vision, a celestial figure appeared who, from that instant on, became her guardian divinity, her protecting spirit. Like Socrates's demon,[32] the celestial figure would serve as Estelle's guide to enlighten her on the nature of her disorder, to direct her daily diet, and to bring her treatment to a successful conclusion.

A friendly exchange began between Estelle and this Angel of Consolation. Here are its main passages: "Ah! Pretty one," said the patient, "do you want to tell me what I need in order to be healed?——Why do you not wish to answer me?——Will you come back every day?——I want you to be completely mine.——You must talk to me a lot——a lot——but I want it to be about my health; otherwise, don't bother me——[33] I would like to fall into a deeper sleep in order to talk to you.——But tell me, pretty one, will I ever again hear the beautiful voices I heard in Peseux?——Am I not lucky to have come to Aix?——I still must thank Monsieur De Castella for having sent me here!!——Without him, I certainly would not be here——For, if he had not lavished so much care on me, you know very well where I would be now!!![34] But, tell me, pretty one, does magnetism suit me?——You do not want to answer me?— Tell me, then, what is your name?——You don't want to answer me?——Ah! My name is Louise-Estelle L'Hardy.[35] ——Ah! Angeline.——What a pretty name! Well, Angeline? Don't you want to tell me the time of your daily visit?——And you say it is because you have been created expressly for me that you are to deal with no one but me?——That is very good.——Am I in a deep enough sleep, Angeline?——Out of ten measures, how far have I gone?——to seven and three-quarters![36]——That's very good.——Farewell, dear Angeline!! good bye!——goodbye!——Kiss me."——(She pantomimed a kiss.)——And the patient woke up at that moment.[37]

Estelle immediately returned to her usual state of alertness. Seeing us laughing out loud, she looked surprised and asked, "What's going on?——What are you laughing about?——."[38]

(In a footnote, Despine explained, "I cannot resist including some textual excerpts with these notes. They were entered daily by Madame L'Hardy. By virtue of this naive depiction—or rendering—of the events, the reader can get an idea of the rapid progress toward healing from the very beginning of the magnetic treatment." Despine allowed the Mother's record to carry much of the story during these seven therapy sessions, but placed them in his endnote 2. For the reader's sense of narrative continuity, that endnote with Madame L'Hardy's transcription was moved into the text here, at its chronological place.)[39]

December 25 On Christmas,[40] after 5 minutes of magnetism, Estelle already sees red grains. They swarm. Monsieur Despine gets her to open her mouth by touching her chin. (*After 8 minutes*) She believes she has half the sleep she had yesterday; her legs jolt and her head drops to her shoulder. (*After 9 minutes*) Grains of fire swarm all the more. She sees red clouds. She sighs and says, "Your fluid is really hot, Sir." (*After 10 minutes*) She can no longer talk to me (it is I, her Mother, writing). "I no longer hear these ladies, except vaguely," she says, "but I hear you very well, Sir——Don't lean on my left leg, it wakes me up." Soliloquy that we do not understand. (*After 14 minutes*) Emilie (her maid)[41] approaches her. Estelle sends Emilie away.[42] "How fortunate that Emilie has gone away," she says. Sigh! There are no more jolts. "Ah! Emilie has hurt me by coming close." Yawns. Big sigh. She hears everything in a confused fashion. She then says, "I am very glad that the oppressive feeling caused by Emilie has passed because it kept me from taking care of myself." (*After 21 minutes*) She has Monsieur Despine place the middle finger of his right hand over her heart and epigastrium[g] without actually touching her; she says she feels a pleasant fluid penetrate her stomach. The finger at its tip appears all red to her. (*After 27 minutes*) A big grain that she had not yet seen appears to her. She asks for long passes from head to toe with quick throws on her face, stopping at the chin. Monsieur Despine's fingers "throw fire" from their tips. "The pretty sparks!" she says. (*After 30 minutes*) "I am worried not to see Angeline arrive." (*After 35 minutes*) Monsieur Despine asks Estelle why she does not speak of her legs. She responds that she is not yet able to do so but that later she will. (*After 40 minutes*) I (her Mother) had Monsieur Despine ask her a question for me. She tells him to say that I am distracting her and that she is busy looking for a way to prolong and deepen her sleep. She asks for passes. (*After 44 minutes*) She sees circles of fire before her eyes and describes various figures. She attributes all of that to the fluid coming from Monsieur Despine's hands. The grains are not like a river, as in the first days, but rather like a swarm. (*After 46 minutes*) Estelle wants an entire pass of throws without stopping anywhere. "Your fluid is really warm, Monsieur Despine." Big sigh. Next she sees an actual magic lantern[43] unroll before her. (She describes different domestic scenes as she goes along.)

Monsieur Despine asks Estelle to what degree she is asleep. She responds, "At eight and three-quarters." She is worried because Angeline is not coming. She asks for throws. "If she doesn't come, I may never heal!!!" Big sigh. During the passes she sees her body and legs, and when Monsieur Despine's hands pass over her she sees festoons of fire. He must stop at her chin and eyes. New vision. It is the garden in Peseux with her cousins and her cat. (*After 1 hour 30 minutes*) She asks for passes and cries, "This fluid is so hot! It penetrates to the very marrow of my bones." (*After 1 hour 35 minutes*) Monsieur Despine puts his watch on her hypogastrium,[g] and she feels sleep deepening. She has felt some fatigue today in her head and stomach, caused by the cooked carrots she ate. "Monsieur Despine," she says, "please put the watch at the pit of my stomach; if it is not there, one eye will close more than the other." (*After 1 hour 40 minutes*) "Is it possible, Angeline, that you are not coming?——Oh, well, if you remain there I will try to heal without you! Tell me, do you enjoy watching poor Monsieur Despine tire himself for me as he does?——I am nine and a quarter asleep. Here I am in Paris, *rue Pavée*.[44] Let's go to the *Luxembourg* gardens——I go by way of *rue St. André*, *rue des Fossés St. Germain*; there is the *Odéon*; let's cross. There is *la Garde*[45] and the violet merchant——Let's go back—*rue Christine*——there is our big living room——That is where my poor Papa died. It causes me too much pain to look——There is Momma and Papa's room——That is where poor Papa got ill; the next room is where my sister and I slept——Poor Blanche, you are in Nantes!!!——(*After 1 hour 50 minutes*) "Please, Monsieur Despine, rub my face with your thumbs.——Hold my chin——Angeline is not coming! Yet, I am in a deeper sleep than yesterday——Please ask Momma not to forget the four-flower tisane."[g] (*After 2 hours*) At Monsieur Despine's request that she take his watch, Estelle moves her hand forward. She likes to touch gold[46] but glass tires her. She does not want anyone to buy her the pocket watch Monsieur Despine proposed; it "costs too much." She puts the watch on her epigastrium,[g] taking care to turn it so that its glass side touches her flannel shirt. Next she crosses her hands on the gold and taps her fingers on the back of the watch as if on a drum or harpsichord. Big sigh. She asks for new passes. In touching the key of the watch, she avoids the steel, since the simultaneous contact with these two metals seems to tire her.[47] Her sleep is good but light. She instructs us that it will be necessary to give her beer tomorrow, if she asks for some. She doesn't seem to want electricity tomorrow. But rather some magnetism? "Yes, Yes."——Above all, it is important to avoid frightening her. (*After 2 hours 15 minutes*) "Ah! Angeline has come."——Angeline approves of the beer, the four-flower tisane and some butter with honey, but absolutely never with salt, despite her demands.——We will also have to give her small handfuls of snow when she wakes up.——"Angeline, should I have electricity tomorrow?" (Estelle listens.) "You believe that it could hurt, but that I'm better off having it done?——Meanwhile, am I free? Why did you not come sooner? You will make me waste my time and you will annoy the good Doctor Despine——Come more quickly tomorrow so that I can question you about my health.——I am going to wake myself up. Goodbye——Goodbye, come back——'Enough passes, Monsieur Despine.'" She wakes up with a start. "My God! Where am I, Momma!!!"

December 26 (*After 3 minutes*) Grains of fire appear, they swarm more than yesterday. (*After 5 minutes*) The swarming increases in intensity and bluish stars appear. Jolts in her arms and legs. Estelle asks Monsieur Despine to hold her thumbs. (*After 8 minutes*) She sees big red clouds. She can no longer talk. (*After 10 minutes.*) Yawns. Big sighs. "Tomorrow it will be necessary to decrease the light coming from the chimney." (*After 15 minutes*) "I no longer want to eat any flan——Momma was saying this morning that magnetism was tiring me——That's not it at all——on the contrary, it is something else that I don't want to say." (*After 20 minutes*) She sings and says "the fluid" makes her sing. If Monsieur Despine stops at her neck while he is doing passes, she says that it wakes her up. (*After 30 minutes*) In a low voice she begs Monsieur Despine to find an excuse to make Emilie leave, for she is tiring Estelle. Emilie leaves. (Estelle's eyes had been closed since the moment the session began, and she had not reopened them.) "All the better," she said, "Emilie has left. My oppressive feelings are disappearing." (*After 40 minutes*) Estelle sees various figures on fire. "Poor Emilie!" She adds, "I fear I have upset her." (*After 45 minutes*) She has a vision with people who sing, but the music is not as beautiful as the voices she heard in Peseux. She asks Monsieur Despine for passes and says, "Monsieur Despine, your fluid has never been so hot!!" She wants a pass around her eyes and she senses that fur would "hurt" her.[48] (*After 48 minutes*) She asks for Monsieur Despine's watch,[49] puts it on herself (its glass side against the flannel), and plays on the back of the gold case, as if on a harpsichord. (*After 52 minutes*) She says, "The gold takes the place of passes over my body,——I want none for the moment except on the face." She puts the watch next to her skin by the epigastrium,[g] but turns it over so that the gold side directly touches her skin, and she rubs the watch in circles at the pit of her stomach. She cries when Monsieur Despine tries to touch her at the epigastrium.[g] Next, she brings the watch to her eyes, rubs her face with it, and puts it against her chin, apparently studying the resulting sensations. (*After 1 hour*) Her arms, almost paralyzed before touching the gold, are now extremely agile.

She sees big blue stars in her head. Moving the watch to the front of her body and placing it on her eyes makes the grains swarm. She asks Monsieur Despine for throws, some of them very rapid. She breathes on the watch and rubs her legs with it. (*After 1 hour 10 minutes*) Then she says, "Angeline is not coming——I will remember this session only vaguely——The glass hurts me to the touch——I definitely do not want a pocket watch, but a 100 franc coin instead——My sleep is good; it is deeper than yesterday's." (*After 1 hour 20 minutes*) "But Angeline is a long time in coming!!! Ever since the watch has been on my stomach Monsieur Despine's fluid works better and my sleep is deeper. When the glass touches me, all the good effects of gold are taken away——Angeline will apparently not come tonight!!! But I wish that Henriette would come tomorrow." Estelle rubs the watch in Monsieur Despine's hand, then sucks on it. She opens her dress, rubs the watch again in Monsieur Despine's hands, and brings it quickly to her stomach, saying that doing so puts her into a better sleep. "Angeline, you think I have endless patience!" She taps her fingers on her knees as if she were playing the piano. A few apparitions follow: an old dead woman, men who frighten her with horrible grimaces, and so forth. Big sigh. She has other more pleasant visions of interior domestic scenes. "Angeline, why are you not coming?——We must continue with the passes.——I see Monsieur Despine's office——He writes, Good God!——so many letters!——He takes a big book. Oh! This bores me! But, look! This one seems interesting. It's not about medicine. There are caricatures;[50] they will amuse me.——He rubs the machine for electricity before I go to it. Oh! That bores me horribly!——I want to stay——Let's go, since we must!——Good! The machine is disappearing." (The session had lasted two hours so far.) "We must continue with the passes, Monsieur Despine, don't you think? Please continue, I would like long passes from head to toe, but slowly, very slowly." (*After 2 hours 10 minutes*) She recites a canticle, another one, then lines from Lamartine[g] (*La Solitude* and *Le Crucifix*). A cantata. Verses from *Athalie*.[g] The Death of the Knights Templar.[51] The strength of her voice, initially weak, increases by degrees and becomes very strong. (*After 2 hours 40 minutes*) She perspires a bit—unusual for her. She asks for more passes. Estelle gets angry that Angeline has not come and that she thus "tires out poor Monsieur Despine." (*After 2 hours 50 minutes*) She launches into another canticle and proudly announces that her memory is getting better. But she's really bothered that she can recite nothing sad, including poetry by Millevoie.[g] While singing, she has been lying on her left side. "Thoughtless Angeline, you have left me this way!!" She then launches into a cheerful tune unknown to me. (Her Mother is recording.) She sings it three times and wakes up after 3 entire hours of magnetic sleep—even more bewildered than yesterday.

December 27 After 2 minutes of passes, Estelle says to Monsieur Despine, "Your fluid is really hot today!!!" She already sees grains of fire. (*After 5 minutes*) Her eyes are heavy, she can neither open them nor talk to me (her Mother). Her arms seem paralyzed. She yawns and sighs, saying she wants Monsieur Despine to stop the passes at her thumbs. (*After 9 minutes*) She sees many large grains. Yawns. Sighs. She repeats her command to stop at her thumbs. "So, Angeline, are you coming or not?——I wish that Mademoiselle Amélie would come in a moment; she does me good. Henriette does not have entirely the same effect on me." (*After 15 minutes*) Estelle's eyes are "nailed shut."[52] "So, Angeline, are you coming or not?" Sigh. She hums a tune, sees a basket of grapes, and regrets that they are not for her. Another sigh. She sings and asks for Monsieur Despine's watch. (*After 19 minutes*) She sees a garden and some pretty cats. But, not for the world does she want any near her. (*After 26 minutes*) "As for me," she says, "I must do all that suits me.——Why aren't Amélie and Emilie coming? If it's all right with Amélie, I would like her to visit first. Momma doesn't have enough patience for magnetism. She always wants me to talk about my health. That will come in due time. Your watch, please, Monsieur Despine."[53] She notices that the glass side is turned toward her; however, as soon as the gold touches her, her arms come back to life. She grabs the watch and rubs her face with it; she then moves it from her chest to her arms, to her shoulders, and then to her legs, all the while tapping her fingers on the bottom of the case as if she were playing the piano. She loosens the straps of her camisole to put the watch on her epigastrium.[g] She rubs the watch on Monsieur Despine's head, then puts it back on her stomach. She requests long passes and throws to her face and says that it will be necessary to magnetize her tomorrow. "Ah! Good!" she says, "there is Mademoiselle Amélie." (*After 40 minutes*) Estelle continues to fidget with the watch. She moves her knees freely without any help. She taps rhythmically on the watch. Each time Monsieur Despine does a pass over her knees, there is more and more spontaneous and visible movement in her legs. (*After 45 minutes*) Estelle takes off her bonnet to rub her head with the watch, saying that it is good for her. (*After 50 minutes*) She sings, puts her bonnet back on, and continues to move her knees. The child sings happily and says vivaciously, "No one should come near me or speak to me, even Monsieur Despine."

She sings a canticle while the watch is pressed against her left cheek. (*After 60 minutes*) She recites Lamartine's[g] *Le Lac*. Since she started chanting (without singing) and reciting, there have been fewer movements in her legs. Next, Estelle recites a fable by Madame Desbordes-Valmore.[g] I (her Mother) upset her quite a bit by saying, "It's odd that there are certain things that she cannot recite." Estelle directs me through Monsieur Despine not to tell her later, when she awakens, that her sleep was not deep enough. Estelle points out that her leg movements are a good omen. (*After 1 hour 15 minutes*) Sometime earlier, she had requested a 100 franc gold coin, not knowing it had been purchased that day in Chambéry and brought here by a delivery man 30 minutes after the magnetic session began. Monsieur Despine puts this new gold coin close to her knees. Estelle lets out a big sigh. "Sleep is complete now," she says. Monsieur Despine puts the gold coin on her hypogastrium.[g] (*After 1 hour 35 minutes*) Estelle asks for a comb and scissors. She cuts a lock of her hair without hesitation—despite the fact that her eyes are closed. She then cuts off another lock of hair and asks for some string. "My good, sweet Angeline, will you come today?" She cuts off another lock. She removes the watch from her left breast and puts it on the other side. She says, "I'll eat some snow this evening and also tomorrow morning. Tomorrow evening, probably, I will rub my legs with some snow,[54] and perhaps I'll crop my hair, too." With considerable dexterity she braids the locks of hair she had cut, even though her eyes were "nailed shut" and hadn't been open since the beginning of the session. She gives one lock to Monsieur Despine. Estelle puts her bonnet back on. She avidly seizes her 100 franc coin, moves it to her stomach and leaves it there, takes back the watch, rubs Monsieur Despine's head with it, and "breathes in the fluid that surrounds it." She self-prescribes beer for the next day. (*After 2 hours*) Estelle asks for some snow. With it, she rubs her hands, her forehead, and her face. She eats some snow and says it doesn't feel cold to her. After eating it all, she says briskly, "Take the saucer away from me, or I'll throw it on the floor." (*After 2 hours 10 minutes*) "I forbid anyone to take my gold coin." Estelle asks for new passes, adding, Now I think that Angeline is going to come soon.

'Is it you? Oh! Finally you are listening to me.——But why did you not come yesterday?——Your excuses always prevail, you know. I am angry that you have come so late.——I have done well to use electricity, haven't I, Angeline?——Was it given to me correctly? You say that they can trust me for the number and the strength of the shocks? I would do well to take them at 1 round, one and a half rounds! If I feel sick I should rub myself with water and vinegar?——and put the watch on my epigastrium?[g] I must drink some beer. I should often ask for water and milk? Snow is good for me? You say that when I am in crisis, I should eat no sugar, and very little sugar in the non-crisis state? Understood. Grapes are good for me? I must be cured this season, and I must eat some of them now, as many as can be found.' " (*After 2 hours 48 minutes*) "I want some passes. 'Goodbye, dear Angeline——come back, come back——' Enough passes, Monsieur Despine," and she wakes up, saying, "My God, where am I?"

December 28 (*After 2 minutes*) Grains of fire.[55] (*After 3 minutes*) She shuts out the daylight and no longer wants me (her Mother) near her. (*After 4 minutes*) Movement in the legs. Estelle says that she is half asleep! Sighs. "You must hold my thumbs, Monsieur Despine." (*After 5 minutes*) "What grains of fire, Monsieur Despine!!!" The night-light tires her even though her eyes are "already nailed shut." (*After 10 minutes*) She asks for Mademoiselle Amélie to come toward her, "but not too close." Estelle regrets that no one has given her any beer today. "It was forgotten!" she adds with a sigh. Emilie always made Estelle tired when Emilie's hand was hurting her a lot. "Aunt Julie tires me a lot but Gramma tires me less than anyone else." (*After 25 minutes*) Estelle is now three-quarters asleep. Monsieur Despine has not stopped holding her thumbs since the moment she asked him to do so. It has immobilized her hands. "If I had a writer at my disposal," she says, "I would have him write a book on polite behavior for these ladies." (We were away from Estelle at the other end of the room, speaking together in low voices; we could hardly hear each other.) Estelle sighs. She notices Monsieur Despine's watch and asks for it. As soon as the watch touches her, her immobilized arms begin to move. Estelle asks for the gold coin and the scissors; she takes off her bonnet. She asks for long passes with throws from her head to her toes. She rubs her face with the watch and is saddened because the watch stops. "We must wind it again," she says. Glass unsettles her. The gold coin is on her chest. "You must bring me my poor Papa's watch with the gold chain." She puts the gold coin on her right eye and the watch on her left eye. (*After 50 minutes*) Estelle says that Monsieur Despine's fluid is red and that the watch placed on the left eye makes that side sleepy. The coin puts her other side to sleep. She sings a Tyrolean song.[g]

(*After 1 hour*) "Aha!——I believe that Angeline will bring me some friends today." Estelle sings a canticle, places the coin on one shoulder and the watch on the other, takes off her bonnet, takes a lock of hair, and tells Monsieur Despine that she doesn't need any help now. She asks for string and paper and says to him, "When you magnetize me, you waste time stopping at the watch or at the gold coin." (*After 1 hour 25 minutes*) She gives a plait of her hair (one she has just braided) to Monsieur Despine. (*After 1 hour 37 minutes*) She asks for whatever she needs to write, but says the nightlight must be turned down "if you want me to see with my eyes closed. You must continue with the passes, Monsieur Despine; your hands don't bother me at all in this whole business." Monsieur Despine tries pressure and insufflation[g] on the pathway of the big nerve trunks (earlier he described these for us) to see whether he can make her legs move. This method works completely, just as he predicted. Estelle asks for some slow passes. (*After 1 hour 50 minutes*) She puts the watch on her legs and rubs her forehead with the gold coin; she has Monsieur Despine blow on them. Estelle taps her fingers on the watch as if on a piano and then enters into a state of quiet and total immobility. (*After 2 hours*) She hums a rather happy tune and believes that Angeline will not be long in coming. Because she remains tired and motionless, Monsieur Despine asks her if there is some new complication with the watch. Estelle responds, "Yes, it is stopped and so I stop, too. The effect this watch has on me is really maddening!!!" (*At 2 hours 5 minutes*) She puts the watch on one side of her groin. She immediately and spontaneously moves her leg on that side. She is deeply asleep and will later remember very vaguely a few aspects of this session. She would like a ring made by her sister. She is less tired than earlier. She points out the people who tire and irritate her; she adds that the same can be said for the timbre of the clock and its "endless vibrations." (*At 2 hours 15 minutes*) "Aha! Dear Angeline, is it you?——'Oh, she is so pretty!——She came so quietly. ('You must continue the passes, Monsieur Despine.') Ah! I heard her——She went away toward the person she asked permission to bring along. Oh! There they are.' Oh!——Zéalida is your name——That's good. Dear Angeline, tell me, must I go for electricity tomorrow? Oh! If it is too cold, may I stay home? (Angeline and Zéalida are friends, both always being of the same mind.) How many turns of the wheel? One, one and a half——two, that would be too strong——and two and a half, much too strong——I decide on the number of shocks?——What caused the crisis that I had today?——Monsieur Despine wants to know, Angeline. Angeline and Zéalida say to Madame B.[56] that the shocks should range from weak to very strong. You say that magnetism is good for me? You say that Monsieur Despine should continue it, but only he——in his absence no one should try it——" (She listens.) Oh! The insufflation[g] is not very useful for the moment——One or two times each session will suffice. I must eat some snow when I wake up, tomorrow too; I must drink some beer, some water and milk often——Cold showers would do more good than hot showers, although these were good in the past. (Zéalida leaves.) Tomorrow, eat lettuce prepared with a light dressing——("You must have Papa's watch brought.") Goodbye, Angeline; bring back Zéalida to me. What? You say that you will bring Pansia back to me? Oh! All the better—— Goodbye! Goodbye!—— "That's enough passes, Monsieur Despine," and she wakes up.

December 29 (*After 3 minutes*) Her eyes are half "shut."[57] "What grains of fire," she says! Estelle feels completely paralyzed except for her mouth; however, she can speak only with Monsieur Despine. (*After 5 minutes*) She is three-quarters asleep. She grumbles and moves to the left when someone other than Monsieur Despine approaches the couch. (*After 8 minutes*) A few sighs. Some jolts. She does not want Monsieur Despine to talk to us.[58] Her head leans on her right shoulder. (*After 12 minutes*) Big sigh. She begins to hum a little tune. She is almost completely asleep, and she wants Mademoiselle Amélie to sing. (*After 15 minutes*) Mademoiselle Amélie must eat no more snow because of her cold. Estelle laughs disdainfully because Mademoiselle Amélie had laughed at the prescription. (*After 20 minutes*) Estelle thinks about her own cure and asks for the watch to end her paralysis. She carries it back and forth from her left to her right ear, and from her left eye to her right one, as well as to her cheeks and chin. She plays the piano on the back of the watch. Estelle asks for the coin and says that the movements of the watch inspire movements in her as well. She moves the watch and the gold coin to her shoulders, to her neck near the clavicle, and then to her groin. As a result, she can move her legs as she did yesterday, however now with greater ease and broader range. (*After 25 minutes*) She asks for her purse in order to write. She moves her hand to her eyes and then to her forehead, appearing to reflect. "Give me some passes, Monsieur Despine." She writes. (*After 35 minutes*) She throws down her paper. Under the influence of the gold coin she extends and bends her legs at will. She hums again and places one hand then the other on her forehead. (*After 40 minutes*) She very much wants to touch Monsieur Despine's gold coins, but not his coins made of any other metal. Monsieur Despine offers them for her to choose from. The silver one stings her so she eagerly grabs the gold one. She loves gold and platinum, without preferring one or the other. She then asks for long passes. (*After 50 minutes*) Monsieur Despine proposes that she put a coin at the nape of her neck and another on her lower back. She does not want to do so today. "You must continue the passes," she says "but very slowly, very slowly." During this time (out of our sight), Estelle herself places the gold coins at the spots indicated by Monsieur Despine and sits up on her own for the first time. A second time she asks for his hands and sits up again. (*After 1 hour 15 minutes*) She takes off her bonnet and makes us leave for 10 minutes. We come back in. Estelle asks for the scissors, some paper, and a cloth. "Angeline," she says, "I hope you will come today; but wait for me to finish what I am doing." (She cuts her hair). "When I wake up I will be angry for cutting my hair——" Monsieur Despine notices that she may stop——"Oh! Whether I get angry or not is of little importance!——my health above all!!! However, I will look awfully funny for I must leave my hair uncut in the back——but you must not tease me, please, for that tires me a lot." (*After 1 hour 40 minutes*) She takes the gold pin from Monsieur Despine's jabot, puts it in her mouth, and rubs her teeth with it. Five minutes later, she asks for some snow and rubs her hands with it. "You must continue the passes," she says. Estelle eats some snow with the gold pin. She rubs the coin with snow and experiences warmth in her hands. She puts her bonnet back on and continues to eat snow. She throws a little snow on all of us, saying, "I am going to prepare myself to greet you well, Angeline; one moment, please." Estelle "sees" and clearly makes out the pin and snow in the plate. Meanwhile, she has not opened her eyes since the beginning of the magnetic session. Her hands feel very warm to her, like Monsieur Despine's very warm hands, despite the fact that her hands are actually very cold to others. (*After 2 hours 1 minute*) Estelle says, "Oh! It's you, my dear Angeline." (She seems to reflect at that moment while leaning on her left shoulder.) "What then!——Zéalida, Elotina, and Pansia, where are they? You did not bring them? Oh! Later, you say. 'You must continue the passes, Monsieur Despine.'" She seems to listen to someone speaking to her, her hand resting on her nose and cheek bone, reflecting very attentively. "Oh! All three of them are there——I love you, dear Angeline——dear Pansia——you have two very nice friends——What are you thinking of telling me, Mademoiselle? I ask you, Pansia,——What must I do tomorrow? Get electricity? How many turns of the wheel? You say, one, and one and a half turns are enough? I am very glad that you will allow me to have snow tomorrow. How much?——All the better, neither too much, nor too little. Some grapes? Yes, and then some beer. I was right to shave my head?——For Emilie's problems, what do you advise Dr. Despine to do? Some laudanum?[g] That's good, very good——cauteries[g] are good too? Yes, and a suitable diet. Back to me, now——you advise me not to repeat the tests with the coin and the watch to exercise my legs. And, why? Please tell me. For fear of fatigue.——Then tomorrow I must eat a bread soup without meat for my dinner.——Angeline, you like the fact that my hair is cut?——That's very good. Goodbye——Goodbye, dear Pansia; goodbye, Angeline, goodbye." Estelle sings for a few minutes, then she says, "Enough passes, Monsieur Despine," and she wakes up.

December 30 (*After 2 minutes*) Grains of fire appear.[59] (*After 3 minutes*) Estelle says to Monsieur Despine, "Your fluid is so warm." (*After 4 minutes*) Her eyes can no longer open. (*After 7 minutes*) She leans her head to the left and lets out a big sigh. (*After 8 minutes*) She is nine-tenths asleep. (*After 10 minutes*) She thinks of something she will tell Monsieur Despine shortly. (*After 12 minutes*) She asks for the watch and, even though her eyes are closed tightly, her hands reach for it with complete accuracy. She takes hold of it eagerly. Estelle moves the watch to her eyes, her cheeks, and her chin. She asks for "slow passes." (*After 25 minutes*) Mademoiselle Amélie comes into the room. "Mademoiselle Amélie," she says, "you will find me ugly, but that is of little importance. I will trim my hair because it is necessary to do so." She flutters her eyelashes and eyelids and says doing so enables her to see light. (*After 30 minutes*) She asks for scissors, paper, and a cloth. She then cuts her hair while she sings. "You must continue the passes, Monsieur Despine. I am very sorry to cause you so much trouble, but when I need passes, I need them." Monsieur Despine tickles her without warning; she does not feel it (*After 35 minutes*) "Momma must say nothing if I cut my hair; I do nothing without a reason, and a good reason, at that. Anyway, Momma can have what I've cut if she likes." Estelle wants to cut her hair on the entire top of her head, that she outlines with her hand. (*After 40 minutes*) A skeleton appears——"Go away, go away," she says. "Monsieur Despine, tell Momma that she must write to Blanche for the rings." (*After 45 minutes*) When Monsieur Despine stops his passes right above a knee, movement in that leg is increasingly evident. "Oh! Sly little hairs, get down——, get down——you are bothering me." (*After 50 minutes*) "I see a big plate of snow——Oh! Now I am deeply asleep." She sings. With the snow, she rubs her head where her hair is cut. (*After 60 minutes*) She puts yet more snow on her head, and she wraps it with a simple scarf. "It must dry slowly," she says, "to have a greater effect." She puts the gold pin in the snow and rubs the top of her head with the pin. She also puts the gold coin and watch in the snow and uses them to rub her legs from her hips down to her knees. As before, spontaneous movements take place in the entire limb.

After 1 hour and 10 minutes of the session, Estelle asks for more snow. She is given a large plate of it. She eats snow with the help of the pin; then she rubs her legs with it. "I need another plate of snow——continue the passes, Monsieur Despine. Oh! It's so hot!——I need two cloth handkerchiefs to wrap my legs in." (*After 1 hour 50 minutes*) She writes me (her Mother) a letter. "Monsieur Despine, don't speak to me while I'm writing to Momma," she says, "that distracts me." After scrubbing herself with the snow, her complexion, which had been so pale, takes on the loveliest coloring. She presses her hand to her forehead and seems to think. "Oh! It tires me too much to write today, Monsieur Despine. Please write as I dictate." Estelle wanted me (her Mother) to write to Blanche for her rings. Estelle wants four rings, each to be sent in a separate letter because, "then I will have news from my little sister four times. She doesn't write me enough——" Intuitively, Estelle understood it would do her good. After all, it benefited her to rub herself with snow. (*After 2 hours 5 minutes*) "Is that you, Angeline? Monsieur Despine, continue the passes; I want to know if Angeline is coming." Monsieur Despine asks her how Angeline is. With a solemn and prophetic tone Estelle responds, "I have seen——I have seen columns and steps surrounded by clouds. Angeline was coming down the steps. She wore a veil as white as snow that fell such that her face could be easily seen. Her long hair was blowing in the wind. A gown as white as her veil flowed to her feet, leaving her arms bare. Her face is lightly colored. Her wide blue eyes are full of sweetness and goodness. She has a pretty nose and mouth, small feet and beautiful hands. All in all, a celestial figure, ever so beautiful——so beautiful and so good that no one has ever—even in their most resplendent dreams——imagined anything like this. Everything about her exudes what she is: An Angel of Consolation." Estelle then says, "I am thinking of you, Angeline. Is that you?" (*After 2 hours 30 minutes*) "Oh! Here they are!——Pansia stayed behind. I did well to rub my legs with snow, didn't I? You say, no electricity tomorrow?——Very well; no electricity. 'Monsieur Despine? Another basin of snow.'" (Zéalida goes to get Pansia. Estelle rubs herself with more snow.) "Oh!——Hello, Zéalida!——Hello, Pansia!——You see me magnetized!!! You approve, don't you?——'They say yes.' And tomorrow, what should I do? More snow and some beer. And for dinner?" (She listens). "Oh! Some buttered macaroni without cheese." (After she rubs her legs, they are as warm as before, if not a bit warmer.) "Pansia doesn't want any electricity tomorrow. You may brush my legs, but no liniment. I will need a little remedy[60] and magnetism, my dear Monsieur Despine; neither my friends nor I can take time away from the treatment; it is too helpful for Estelle——However, you have my permission to go to Annecy to see your family and celebrate Twelfth Night." (*After 2 hours 45 minutes*) Estelle finishes rubbing her legs with gold. "I want to rub my head with snow," she says next, "as that always makes me feel better, doesn't it, Angeline?"

Monsieur Despine explains to us that the essence of magnetism and its therapeutic potential are still little known. Estelle quickly interjects, "Oh! Yes, unfortunately," and sighs deeply. She continues to rub her head and hands with snow. Estelle then asks Angeline what she should eat when she wakes up. "A baked apple?——Yes, good, a baked apple." (*After 2 hours 55 minutes*) "More passes, Monsieur Despine. 'What do you say, dear friends, about insufflation?[g] Monsieur Despine can do it.'" Monsieur Despine blows three times on each knee, exhaling as long as he can. The effect is very noticeable on the corresponding leg. "You must warm my bed and make sure that I'm not cold when I wake up" (from magnetic sleep). (*After 3 hours 5 minutes*) "Monsieur Despine, magnetize my chocolate, please, it will be better."——(*After 3 hours 12 minutes*) "I see a red cloud.——'Dear Angeline, I will follow all your commands.——Goodbye to all——come back, come back tomorrow,'" she says eagerly. Estelle kisses Monsieur Despine three times, hums for a moment. "Enough passes," she says and wakes up immediately.

During electrification today Estelle had two crises, clearly brought on by electricity. The first lasted fairly long. When asked why that had happened, she answered, "Because the watch stopped." The watch had indeed stopped. Monsieur Despine had done so intentionally, setting it for only a defined length of time to see if he would observe any phenomenon in Estelle that he had observed in a few other cataleptic patients. She had him remove the small, yellow copper chainlets touching her feet, since they tired her. In the first crisis, he noticed more lipothymy[g] than spasms. In the second crisis, he noticed the reverse—more spasms than lipothymy[g]—leaving her legs completely stiff. She begged Monsieur Despine to relax them for her; he did this immediately by pressing his finger along her nerves. By pressing under one ankle, that foot immediately relaxed; by then pressing under the hamstring on the same side, her leg also loosened up. Finally, by pressing on her groin, her thigh relaxed like the rest. Estelle could then, at will, bend and stretch the entire pelvic extremity on that side. The other side, however, remained rigid until Monsieur Despine relaxed the entire leg all at once by pressing with one finger in the fold of her groin. He explained that these phenomena were a function of electromagnetic action on the nerve trunk as well as on the entire region served by it.

At other times, Monsieur Despine relaxed both legs by simultaneously pressing on the two inguinal[g] plexi. We had seen him do this before. In one hand he would hold a foot by the tips of the toes; in the other hand, he would hold the other foot in the same way. In most interventions like this one, both legs relaxed progressively and rather quickly. Monsieur Despine calls this technique "acting by means of the galvanic circle."[61]

December 31, 1836 until Early Morning, January 1, 1837 After one minute of magnetization, grains of fire already appear.[62] (*After 2 minutes*) Estelle's hands are paralyzed. She has her shawl taken off her since "she is too warm." (*After 4 minutes*) She is half asleep. (*After 8 minutes*) She says to Monsieur Despine, "Sir, since yesterday I've been noticing something on you that tires me; I don't know what it is——but, I am certain that it wasn't here before.——" In fact, yesterday and today, Monsieur Despine carried a little 10-centimeter magnetized bar in his side pocket to perform electromagnetic experiments on Estelle, similar to those he had tried on other patients. He had not mentioned those experiments to us. A few moments later, he put the bar aside, far from the patient. From this point on, Estelle no longer complains of anything. (*After 10 minutes*) Estelle asks for some towels so that she does not have to lie directly on wool. The square part of the watch key[63] touches a finger and causes her discomfort by "burning" her. (*After 11 minutes*) She is three-quarters asleep. She undoes her bonnet, moves her watch to her forehead, then to her head; she places the 100 franc coin at her sinciput.[g] (*After 14 minutes*) She notices that the point where two different metals join tires her by giving her a kind of electric shock. In just a quarter more she will be completely asleep. (*After 18 minutes*) Estelle requests scissors and some snow. She again cuts her hair, and there is much more movement in her legs. She cuts a lock of hair on her left side and two locks on her right, without wanting to tell us why. (*After 24 minutes*) Estelle pushes her sleeves back and rubs her arms with snow. Next she rubs her head with it. (*After 33 minutes*) She places her 100 franc coin on the top of her head and plays with the watch. She informs us that she will be able to eat snow without sugar from now on whenever she needs to be cured of a stomach ache. (At this moment the temperature was 5° C.) (*After 47 minutes*) Estelle asks for another plate of snow and takes the gold pin from Monsieur Despine in order to eat with it. She puts snow on her sinciput, her forehead, her nose, and her mouth. Hiccoughs, which she had since the evening before, stopped while she applied the snow to her body and as she ate it. This was especially true when she also put the gold watch, the pin, and the 100 franc coin at the base of her back.

After one hour of magnetization, Estelle sings. She interrupts herself, because she says she is sad not to be home for the New Year. (*After 1 hour 7 minutes*) She had prepared some candies for Monsieur Despine and she insists that he take them. (*After 1 hour 10 minutes*) She vigorously rubs her legs.Then she says, "Monsieur Despine, you must continue your passes. Long passes from head to toe put me to sleep and cause a sweet, calming liquid to enter my body." She rubs her legs with the watch, making sure to use the gold side. (*After 1 hour 30 minutes*) She rests the watch on one side of her groin and the 100 franc coin on the other side. She appears to be thinking. Then, five minutes later SHE MOVES BOTH HER LEGS OUT OF THE BED ON HER OWN . . . SHE SITS UP. . . . When we least expect it, SHE STANDS STRAIGHT UP AND BEGS MONSIEUR DESPINE TO EXTEND HIS HAND TO HELP HER WALK. We are all stunned. We dare not believe our eyes because, until now, all magnetic sessions and somnambulistic experiences have taken place in bed without her ever leaving it. Estelle steps forward and comes close to me (her Mother). She kisses me tenderly and kisses Monsieur Despine. Then she says to me somewhat maliciously, "And so, Momma! Do you still fear that the snow is harming me?——" She returns to her bed and soon gets out again, asking Monsieur Despine to support her with the end of his finger. NEXT SHE WALKS ACROSS THE ROOM AND LOOKS IN THE ALCOVE FOR A PLATE SHE HAD PREPARED WITH MY NEW YEAR'S GIFTS. Her eyes have been completely closed since the fourth minute of this session. Moreover, the apartment had only an ordinary nightlight hidden behind a screen. It illuminated only the place where I was transcribing all the details of the sessions as they unfolded.

After getting the candy and fruit plate (including grapes), Estelle walks again across the apartment to present it to me in my little corner and to wish me a Happy New Year. She goes back for another plate and carries it to Emilie (her maid) who was already in bed in a large room next door. Estelle approaches the bed and says, with the same spiteful and cunning tone, "And so! Emilie, is the magnetism still harming me? Do you still find it dumb?——Am I still pretending?——Oh well, no hard feelings, take your New Year's gifts."[64] Estelle dances as she leaves Emilie's bedroom, barely supporting herself on Monsieur

Despine's finger. She returns to her own bed. Then, in an affectionate tone, she thanks Monsieur Despine for all the attention and care he has given her and tells him how happy she is to have come to Aix. Estelle assures him that she will be grateful for everything and will always remember him.

"You are so good, so patient," she says; "I love you so much; oh yes, I love you so much!!! But we are not yet finished," she adds. "When I wake up in a little while, I will still be paralyzed. But we will not regress. My daily somnambulistic practices will strengthen my arms and legs; magnetism will give them life, and my natural instinct will do the rest——but I beg you, Monsieur Despine, please keep the others from disturbing me. The upset harms me a lot and may greatly prolong my illness—perhaps even make it worse."

(*January 1, 1837*) At midnight, Estelle again gets out of her bed without asking for help, EVEN from Monsieur Despine. She goes to the pantry, brings out a plate of meringues and eats half of one. She wants Monsieur Despine to eat a whole meringue, and then offers the rest to all of us. She walks around her couch, using only a chair to hold herself up. If anyone tries to help her, she gets angry and says, "Let me do it, I know what I need." Then she sits down on my lap and wants Monsieur Despine to sit close to her. Estelle asks for some snow and again rubs her legs with it. She often repeats to me somewhat mischievously, "And so, Momma, do you still believe the snow is giving me a cold?" With the same naughty attitude, she throws snow at some of us in the room. Afterward she expresses regret for having been so foolish.

Estelle hugs me again, and kisses Monsieur Despine again. If she touched me on the face during these tender caresses, I "burned" her. She would immediately take Monsieur Despine's hand in hers and pass it over the "burned" area as if to undo the painful feeling it gave her.

Eventually, Estelle cheers up, speaking energetically, telling us very funny things, and teasing all of us without losing any of her earnestness. When asked how her actions are so precise although her eyes are closed, she says, "But I sense everything. Dr. Despine's son may now come when he wants. I believe that he will no longer say that I am merely pretending to get strong."[65]

Estelle wants to lie down on her bed. She drops her 100 franc coin and immediately loses strength, falling to the ground. As soon as the coin is retrieved, she grabs it again, gets herself up, and walks across the room without any support other than the tip of Monsieur Despine's finger. She asks us to "make a chain" in order to increase

Photo 7 Portrait of Constant Despine (1807–1873) Antoine Despine's son.

Source: From the private archives of Philippe Despine, Dijon.

her strength.[66] Next, she goes back to her bed. She seems happy. And me! (This is still Madame L'Hardy writing.) And me!!! Even more so! I am convinced that this poor little invalid may finally be cured through care, patience, and the necessary perseverance. All that we have just witnessed could not be caused by an organic illness of the spinal cord or by an illness of the solid or soft parts that envelop or encase it. God willing, Monsieur Despine will be rewarded a hundredfold for the good he has done me. His astute and keen observations uncovered the nature of this strange illness. His persistence surmounted all obstacles. His experience increased our hope for Estelle's cure. Above all, he correctly identified the most effective treatment method for an illness that had confounded so many in the medical establishment.

(*At 2:15 in the morning*) Estelle asks for more long passes from Monsieur Despine. "They bring warmth into my body," she says. A moment later, she cries out, "Is that you, Angeline?" She remains pensive and silent for a moment. "Ha! It's you, Pansia and Zéalida; then where is Angeline? 'Good. They are going to look for her.' " (Zéalida remains.) "Tell me, my dear Zéalida, tell me, I beg of you——what do you think of the use of gold and magnetic fluid? These seem to give power to a little bar of iron!!! It's my own personal magnet." (Angeline arrives.) "Angeline, isn't Monsieur Despine wonderful?——'Momma? You must write for the rings and for Papa's watch.——Tonight's session is a good omen.——Today I must eat sweetened meatballs for dinner——milk and beer——and if I still have indigestion you must put a big gold coin on my stomach and rub my forehead with snow——Tomorrow, complete rest, except for magnetism.——Oh! I would just love to have a pretty little bird!' You are leaving me already, dear friends? It is too soon!!! Goodbye——Goodbye——Goodbye——'Enough passes, Monsieur Despine.' " She wakes up.

The Mother's transcription ends here. Below Despine generalized about this Phase One: it includes about 20 more days in January.

Somnambulism appears so extraordinary even to those who are familiar with mesmerism. It is thus understandable that those who are unfamiliar with somnambulistic states consider them to be unlikely or unbelievable.

Starting on December 22, 1836, then, Estelle was magnetized every evening through January 4, 1837. At that point I left Aix,[67] so Estelle had no magnetism between January 4 and January 7. Magnetism resumed the evening of January 8, though, and continued through the 22nd. On that date Estelle was involved in a minor domestic misunderstanding, resulting in her lapsing into a spontaneous crisis of somnambulism. From that time on, she neither needed me to induce magnetism to enter a crisis state nor needed to pass successively through all its trance states. On December 24 somnambulism was complete in Estelle for the first time.[68] Every evening thereafter, when in crisis, our young invalid would self-prescribe her diet and her treatment modalities for the next day. She never gave these recommendations without consulting her tutelary spirit, Angeline. We always took precise notes of her recommendations, since the patient was amnesic once she came out of the crisis state. The treatment plan consisted in letting her eat more or less everything she asked for. It also followed some of Estelle's directives, such as to continue the magnetism, let her cut her hair and wash her head and legs with snow, take surprise showers and swim baths (when the time would come for them), eat snow, suck on ice, drink beer instead of wine, drink a lot of milk, avoid meat and fatty broth, drink coffee and red currant syrup, and sometimes take Garus's Elixir[g] straight or diluted with water. We were also to use electricity in all its forms. Estelle instructed us that she never be contradicted or bothered while resting in a good, tranquil restorative sleep. Finally, she demanded that she be allowed to do whatever she wished, "a right," Estelle said, "that I will never abuse."[69]

In Phase One, three core groups of phenomena emerged:

General Symptoms I noticed a sensation of a real fluid leaving my fingers and acting on the patient in a very obvious way. Estelle experienced fantastical visions, heightened sensitivity of the five senses, and an appetite for snow. She bathed and enjoyed herself in an icy bath during somnambulism, but out of crisis she could manage only if covered in eiderdown and quilts. Certain people, including her Mother who was closest to her, disgusted her when she was in crisis.[70] She felt sympathy from some people and was drawn to them; she felt antipathy from others and was repulsed by them. Estelle perceived these interpersonal forces as influencing her to varying degrees. During a crisis, there was remarkable development of her intelligence, memory, imagination, and all her physical and emotional faculties. She would then return to her ordinary state after a fit (crisis). Estelle would indicate daily what she needed. She was concerned only for herself; she dealt with people only in so far as they related to her. Otherwise, she interacted with others to be polite but always with repugnance and fatigue, since it was primarily her own well being that motivated all her actions. She chose the types of magnetic passes to impact her paralyzed limbs as well as everything concerning her diet and hygiene. When in crisis she responded in very unusual ways to the galvanic action of metals.

Phenomena of Locomotion By December 22, 1836, there were still no spontaneous movements in Estelle's legs. On the 24th, they moved slightly under the influence of magnetic passes. On December 25 and 26, these movements showed a gradual but barely noticeable increase. On the 28th, after rubbing the pelvic extremities with snow, Estelle made some spontaneous movements that demanded the greatest muscular strength. Indeed, she sat on her bed without feeling the slightest pain or fatigue in her back. She stretched her legs out in front of herself, between 45 and 50 centimeters above the surface where she was sitting. She then shook her legs in the air and bent her knees effortlessly in all possible ways.

On December 29, the galvanic influence of the metals (or the electrometallic force) became more apparent, and the movements in her leg muscles increased very noticeably. On the 30th, Estelle finished cutting her hair, especially on the top of her head, which she then covered with snow. On the 31st, she stood and walked on her own. From January 1, 1837 until January 12, Estelle's movements would stop when a gold watch she held by a Venetian chain[71] above herself would stop. She remarked that a watch that worked badly tired her greatly and that silk, glass, and Spanish wax "burned" to the touch. She added that the square key to my watch also "burned," particularly where the steel part joined the gold. When Estelle's arm or leg accidentally touched a yellow copper pin, that limb became rigid, whereas a gold pin relaxed it instantaneously. The slightest delay in following her somnambulistic instructions increased the intensity of her crises, her muscle tenseness, and her irascibility. A fit of hiccoughs disappeared as if by magic simply by applying pressure along the pathways of the eighth cranial nerve. This pressure could be applied by any two fingers—two fingers from the same hand or by one finger from each hand. Simply blowing on a nerve plexus connected to the contracted limb relaxed all parts that received life and movement from this neural network; movement then returned to that limb. A gold coin placed on a nerve or its neural network increased the muscular strength of every limb connected to that nerve bundle (see my endnote 3).

Progressive Development of Phenomena Manifested during Somnambulism, Its Marvels and Its Pitfalls From January 12 to 22, the general and locomotive phenomena continued to manifest themselves and improve by the day. Estelle prescribed that, when in crisis, I not take her out of her state of contemplation or syncope since, she would say, "I am taking care of my healing." A watch key made completely of iron or copper bothered her less than a key made of two different metals (either juxtaposed or soldered together). She felt a striking difference when she touched glass or clay pipes or porcelain. Glass "burned," clay pipes seemed cold to her, and porcelain felt lukewarm.[72] Rissoles, anything fried, all meat and animal products (except milk and eggs), but especially grease and oils, gave her heartburn. On January 17, because she felt provoked, Estelle had a fit of rage for the first time. Two days later, the first phenomena of catalepsy appeared in her legs, and I was able to put her to sleep magnetically by blowing for a minute on her eyes. Feathers tired her; touching fur stiffened her, but gold placed on a stiffened limb relaxed her muscular spasms. Thus, a few rubdowns with a gold coin quickly repaired the harm done by feathers, silk, furs and other insulated or idio-electric bodies. The mere presence of the color poppy red as well as shades of poppy or violet put Estelle into a crisis.

In my notes and in the Mother's transcription, there are some very remarkable entries that capture the spirit of magnetic somnambulism. Treating more than 20 cases of this malady during the past 15 or 20 years, I have consistently observed similar magnetic phenomena. This particular state of mind can be characterized by elevation of thought, apt choice of expression, precise articulation of an idea, and prompt judgment. Even though influenced by circumstance and education, these phenomena are so obvious and predictable in all patients with this disorder that a physician who has seen and studied one case would recognize it in another patient.

In Estelle, as in all patients like her, I saw an absolute independence of thought and the most inflexible of wills. This rigidity undoubtedly comes from the quickness of somnambulists' judgment, which results from their very extraordinary heightening of intelligence during this state and their ability to embrace the past, present, and future simultaneously for everything that concerns them personally. They exhibit extraordinary irascibility when challenged because they cannot imagine that other people around them do not see as well and as clearly as they do. Merely questioning their inflexible will can insult them. In this state the mischievous spirit manifests itself by making the patient try to find fault with all who want to tease her or find fault with her.

Phase Two of the Magnetic Cure in Aix-en-Savoie: January 22, 1837–June 30, 1837

This Phase Two of Estelle L'Hardy's magnetic treatment extends from January 22 to June 30, 1837 when she left Aix to return to Neuchâtel. The following section describes the progressive unfolding of Estelle's unusual nervous symptoms.

As stated earlier, Estelle experienced her first noteworthy crisis of spontaneous somnambulism on January 22, immediately following a small domestic quarrel. This somnambulistic crisis was not induced by a planned magnetic intervention. From that moment on, she could open her eyes while somnambulistic; until then, they had been "glued shut" during the somnambulistic state. As long as she was in this trance state, she could take long carriage rides, stand up, and even run around the entire day without the slightest fatigue.

On January 28, the invalid lapsed spontaneously into a crisis state having once again been involved in a domestic quarrel. Later, before leaving this magnetic sleep, she experienced three jolts throughout her entire body for the very first time. When questioned about this new phenomenon, Estelle responded that, with each jolt, there was a jump[73] within her from the preceding state toward the one to which she was headed--either from the somnambulistic crisis state to the waking state or from the alert state to somnambulism. On her own, she decided to try reading with her fingers. Her attempt was inconclusive. Having her eyes closed (or "nailed shut," as she would say), she could distinctly see everything going on in her room. At times, she could even specify what was happening in the room next door!

On February 3, she took an eight-hour carriage ride and behaved as if she had never been sick. She visited friends in and around Aix and Chambéry. Her friends were greatly surprised because they had seen her lying in bed only a few days earlier. They knew that, while in her natural waking state during at least the last six months in Aix, Estelle could not put a foot on the ground or take a step without hurting herself (see my endnote 4).[74] The phenomena of catalepsy increase by the day.

On February 5, a new domestic disagreement caused her to have yet another fit of rage. Estelle calmed down only when we reacted with silence and faced her acts of fury and irascibility with complete composure.

On February 9, Estelle made another trip to Chambéry. She walked about town for more than six hours to see the sights, visit people, and run errands with her Mother at different shops. During this six-hour walk she observed everything. She would sometimes converse in a manner beyond her age and would display behaviors that were far more mature and complex than those in her natural waking state. When cats walked by her, she misperceived them to be on fire and immediately lapsed into catalepsy. That same day, after seeing a fur tippet on a woman from Neuchâtel who happened to be in Aix, Estelle completely stiffened her arms and legs. A few magnetic passes started to shift her out of this state. The stiffness disappeared completely by rubbing her rigid limbs either with a 100 franc coin or with a gold watch.

On February 12, she "heard through her wrist" even though we were whispering. That same day, she began the Scottish Showers. She walked to and from the baths, saying that the portable chair would hurt her. From this time on, each day was divided between 12 hours of somnambulism and 12 hours of bed rest. On February 21, our patient exhibited phenomena of imitation for the first time (see my endnote 5). Beginning on the 22nd, the crises increased in number and were divided among four different, distinct states: active crisis, "dead" somnambulism or syncope, catalepsy or the cataleptic state, and imitation. This last form of crisis, as manifested by our patient, included the various phenomena of echo, attraction, and repulsion (see my endnote 6).

The therapeutic modalities used during Phase Two of the magnetic cure were limited to massages with snow; foot baths in a nearby brook fed by mountain water; Scottish Showers; very light electrification Estelle could administer to herself with the machine; and finally, a vegetarian diet Estelle self-prescribed daily. During the crisis, exercise lasted without interruption for about 12 to 14 hours. When Estelle could run in the snow and flop down into it, she would come back with her slippers and skirts covered with tiny ice and snow pellets, as on the coat of water spaniels and long-haired dogs who have romped in the snow. At these times, she experienced complete joy and well being. She would often bury herself in the powdery snow or eat spoonfuls of it with great delight; she did the same with ice that she preferred in the form of icicles that hang from roof gutters after a night's frost. When she would feel the blood rushing to her head, she would run to the street, sit down in the snow, and pack some on her head in the shape of a cone.

Estelle's diet was remarkable, both the one she followed when in the crisis state and the one she felt forced to follow during the waking state. During a crisis, when she returned to her former good health and health habits (see my endnote 7), Estelle had an appetite for the nutritious foods she had liked in her younger years. Moreover, she would eat abundantly with impunity and without incident. During her natural waking state, however, she could not deviate from her usual diet of vegetables, milk, and eggs without getting cramps, indigestion, or nausea, or without developing other problems that had severely weakened her in the past three years. Moreover, she would pay dearly for the least self-indulgence and for eating what others persuaded her to eat when their tastes were different from hers.

It was as if Estelle had two stomachs, one for the crisis state and the other for the waking state. Even more unusual, the food eaten abundantly during a crisis did not satiate her in the waking state, and vice-versa. In fact, when Estelle moved from one crisis state to the other, she would experience the appetite and the needs aligned with her newly awakened physiological state. How could the digestive tract operate in this unusual way? What happened to the food that had satisfied her great hunger during the crisis state? Where did her preferences for certain foods go? She had preferred cold beer, weak coffee, snow and ice, almond milk, vinegar, and red currant syrup. A few minutes later she favored hot beverages like weak herbal teas, coffee with milk, milk heavily diluted with water, sugar water, and four-flower tisane. For a long time, her daily diet had consisted of those drinks, some garden vegetables (including asparagus, lettuce, chicory, wild salsify, sorrel), and soft boiled or scrambled eggs with cream. How was it that her food preferences changed the moment she switched from the sleeping to the waking state and vice-versa? Most assuredly, recently ingested food could not have been sufficiently digested, so as to trigger hunger again. I leave this question to physiologists more gifted than I. Despite my lack of understanding, I assert the truth of this phenomenon. Later, we will see how, as her cure progressed, the preferences of these two states gradually blended in Estelle.

These curious observations have often led me to draw a comparison between abnormal pathological states and the nature of electricity. It is only when positive and negative electricity remain separate from each other and then are forced in motion "of departure or opposition" by an electric generator that electricity can manifest itself in a measurable and noteworthy way.[75] In a parallel manner, it is possible that the normal or natural state of health results from the alternating of two analogous principles (positive and negative) that constitute animal electricity. In this parallel between electricity and animal magnetism, the brain represents the center, the nerves the conductors, and the nervous fluid the vehicle. Perhaps electricity and animal magnetism function according to identical principles. Among these principles, "separation" and "departure" of magnetic functions might constitute the primary cause of all illnesses of the nervous system.[76]

In any case, the lifestyle changes observed in Estelle are compelling enough for the history of nervous illnesses to interest medical professionals. These new types of observations should incite physicians to be alert to these sorts of phenomenologies in their patients. Estelle's disorder is certainly not unique in the annals of medicine.[77] What I have observed, I have keenly observed; I dare to assert its truth. Others can observe these symptoms as well as I can. Both the inferences drawn from my observations and the basic phenomena themselves can serve further investigations either as a starting point or as a rallying point.[78]

I now continue the account of Phase Two of Estelle's treatment with magnetism, electricity, and the springs. I worried little that my observations and research met with disapproval, even mockery. Rumors spread that I had gone mad or that I was more consumed by ambition than by a commitment to scientific advancement. Clearly, scientific progress must inspire physicians if their work is to profit humanity (see my endnote 8).

On March 1, I left Aix for a few days. Estelle had predicted that, after my departure, she would be very sick and that at times she would become irrational. No one was to bother her or worry about her. In fact, on March 4 she had numerous hallucinations and was irrational for several days. No one in the house could communicate with her. Even her Mother, whom Estelle adored during the waking state, "burned" her as soon as she was in the crisis state.[79] This antipathy was even stronger on those particular days in March. Since no one could reach her, they had to give up and remain silent spectators to all her extravagances.

On March 8, Estelle's body was covered with red stretch marks. On the next day, while sleepwalking, she recounted the entire story of her illness through January 1, 1835. She spoke slowly but constantly for 40 minutes, apparently absorbed in deep reflection. There was a prophetic tone in her voice; she used expressions she was unable to use during the waking state. On March 13, the color of her urine darkened and its odor became very pungent. On March 16 and 17, she bathed her legs and feet in a brook fed by mountain snow. She enjoyed these 40- to 45-minute baths. Her typically pale skin took on a crimson hue. All her veins, invisible on the surface of her skin during the waking state, were now completely visible. On March 20 and 21, Estelle resumed telling the story of her illness; I wrote as she dictated. She described the events with such an exceptional order and method that I myself might not have been able to do as well on a first try.

Between March 22 and 25, our little patient had panic attacks accompanied by great emotionality. Several times, she wanted to bite; she attempted to satisfy this desire on the first person to pass by, namely, on me. In her excessive physical and emotional excitation, she frequently demonstrated the phenomena of imitation and of echo (see my endnote 9). These behaviors had been triggered by her unexpectedly seeing a poppy-red shawl. They were further aggravated by the screech of a knife cutting through hard, moldy bread.

On the evening of March 25, Estelle talked to us about her "ball" (see my endnote 10). She told us that it had been rapidly growing in size the last few days and that when it would burst, she would immediately feel better. She predicted that her health would continue to improve from that point on. She also predicted that she would be able to take a few steps without help—when she was not in the crisis state.

On April 1, Estelle amused herself by playing April Fool's tricks on everyone in the house. On the third of April, I was again absent. Our young patient took advantage of my absence to overindulge in Garus's Elixir,[g] which was usually diluted and administered in small doses. She proclaimed that this liqueur "lit her stomach on fire"[80] and gave her ecstatic fits of delirium, impatience and rage, as well as excessive pain in her leg muscles. A red coat, seen by chance, caused her dreadful anxiety. Nothing calmed her except a music box,[81] silence, darkness, and the composure of those around her.

On April 14, I returned from Annecy. Estelle's "ball burst" and she walked for the first time out of the somnambulistic crisis. While in much pain, she took 10 or 12 steps; her gait was unsteady. Afterward, she was unwell. We had to sit her on a chair and then carry her to bed. But she had walked while in a perfect waking state! This alone was enough to instill hope in Estelle's beloved Mother and to motivate her to push this extraordinary treatment to its conclusion. Between April 15 and 30, Estelle in crisis no longer needed as much gold in order to walk. She traveled to Chambéry without carrying all her gold coins, managing to sit up straight in the carriage without any discomfort.[82] Meanwhile, her Mother had put additional coins into her own pocket. Estelle got out of the carriage, forgetting to ask for them, and she even walked for quite a while. Her Mother, however, noticed that without the gold coins Estelle's gait was a bit slower. A few times Estelle came to ask for her Mother's hand,[83] at which point her stride became quicker, more confident, and less tentative. When Madame

L'Hardy would secretly give the gold coins to me or to someone else in the group, Estelle would weaken immediately, get impatient about not being able to go more quickly, and complain of fatigue. Finally, she would collapse.

Despite the noteworthy improvement observed during these 15 days, Estelle had several hallucinations.[84] One day, when I meowed like a cat, she actually mistook me for one. This reoccurred when I imitated other animal sounds such as a sheep's and a rooster's. On April 22, Estelle had a 45-minute bout of uncontrollable laughing that she made no attempt to contain. In fact, she seemed to be completely unaware of its happening.

On April 27, she took her first dip in the pool. The water was at 21° C, and Estelle seemed delighted to be standing there up to her knees, but the water was too cold for her to go any deeper. The temperature of the water in the pool was both variable and layered. Estelle went in up to her waist only when the temperature of the upper layer of water (about 30 centimeters deep) had been raised from 26° to 28° C. Therefore, the temperature of the water at the bottom of the pool remained constant at 21° to 22° C. Estelle continually sought the coldest spot of the lower layer and the warmest spot of the upper layer (that is, closest to 28° C).

Photo 8 Nineteenth-century representation of the Baths.

Source: Courtesy *Archives des Thermes nationaux d'Aix-les-Bains*.

While thus wading and searching for the spot that best suited her, she came to the middle of the pool where the water was just above her navel. Suddenly, Estelle became cataleptic, standing like a statue. Her Mother became anxious. The lifeguard was absent. If the child were to fall in the water, we could do nothing but jump in to rescue her. Meanwhile, since Estelle was not losing her balance, I was able to reassure Madame L'Hardy. I called to Estelle but there was no response. I knew of no other way to communicate with her. Although unsure of what to do, I was quite confident that the instinct that had led Estelle to her chosen spot in the pool would also guide her out. Nevertheless, I could see that Madame L'Hardy's anxiety continued to increase. She had already sacrificed so much for her daughter who had seemed so close to a cure. Here is what happened.

Standing in water one meter deep, Estelle—in the middle of this six and a half by fourteen and a half meter pool—could be approached only by two ramps, one at each end of the pool. She was staring at the wall opposite the wall where her Mother and I were sitting. I called her several more times but got no response. I walked around the pool to call her again. My voice was practically inaudible because of the high ceiling of this large room. I gestured toward Estelle, who looked like a statue. I did some magnetic passes from afar, but

Photo 9 Drawing of the pool where Estelle most likely waded + Swam.
Source: Courtesy *Archives des Thermes nationaux d'Aix-les-Bains*.

Estelle remained immobile. Since she was neither moving nor losing her balance, I sat by Madame L'Hardy on the first step of the ramp taken by Estelle and continued to reassure her. We were both very determined to wait for whatever might come and not worry excessively. While sitting there, I suddenly remembered that water carries voices well from one side of a lake or pond to the other. Applying this principle to animal magnetism, I dipped the index finger of one hand into the pool while saying Estelle's name. The "statue of Pygmalion" immediately came to life. Although she had been cataleptic, she did a pirouette, turned to face us, and came straight to me. From that moment on, Estelle spent between one and two hours in the pool almost every day, trying to swim with safety equipment while accompanied by the bathing attendant. She walked for the entire day while in a somnambulistic state, and in the evening she tried to take a few steps while in the waking state.[85]

On May 7, we left to spend a couple of weeks in the Bauges Mountains[86] where I had business. I escorted Estelle and her Mother there. We made this nine-hour trip partly by carriage and partly by foot. The patient endured it as comfortably as the other passengers. When we were passing through a town, her face flushed and became animated. When she saw a fountain, she asked to stop the carriage, stepped out, and ran to it. During a good quarter of an hour she splashed cold water over her head to wash and refresh her face. She drank some water mixed with beer and then climbed back into the coach. She seemed to enjoy the last part of the trip, showing interest in all she saw. She compared the beautiful views of the Chambéry Basin and the Valley of Isère with views of Switzerland. She sang, chatted, and recited verses and amusing stories that reminded her of childhood, including elementary school.

They arrived in the mountains. While Estelle was in the crisis state there, she thought only of picking flowers and running around from about 10:00 in the morning until about 10:00 at night. During the stay in the mountains, a few unexpected and upsetting events may have slowed her progress. First, she lost her 100 franc coin, a sort of talisman for Estelle. Indeed, during the preceding five months, she had been unable to take a step without it. Although everyone else knew of this loss, I was initially unaware of it. Estelle was particularly affected by the loss, as she was loathe to disappoint her Mother whose sacrifices Estelle recognized and deeply appreciated. She wanted to direct the search for the coin in secret, but it was not immediately found. Finally, noticing that Estelle was behaving strangely, I feared problems and insisted on knowing what was going on. As soon as I knew the reason for Estelle's behavior, the coin was replaced with one like it, and everything returned to normal.[87]

Estelle would very often become cataleptic when the house cats scampered between her legs at the dinner table or rubbed against her feet while she was sitting or walking around the apartment. She also experienced occasional "running crises": suddenly she would have an irresistible urge to run that could last from 5 to 15 minutes. One day while in crisis, Estelle was startled when she saw a child her age stricken with epilepsy, having seizures with countless gesticulations. She had never seen this nervous condition before. I wanted her Mother to distract Estelle by taking her out of the room, but Estelle insisted on staying, apparently interested in this patient and in my consultation. Suddenly, she herself was struck with epileptic-like behaviors, imitating the spasmodic movements of her unfortunate companion. Estelle was soon rid of these behaviors when we distracted her, gave her a footbath in cold water, and took her for a walk.

Overall, the stay in the Bauges significantly improved Estelle's health. On May 26,[88] we left for Annecy where Madame L'Hardy and Estelle had family they wanted to meet. We had almost 30 kilometers to cover. Two-thirds of the journey was goat paths rather than roads. To make this trip, we were forced to ride a mule or horse, the only transportation available. Estelle's condition posed a problem for such a trip. There were no guarantees that the crisis state would be sustained; she could shift to her waking state several times during this long trip. Then, despite precautions and directives given to the guide, she could fall to the ground because she was always lopsided and still greatly weakened. Encouraging Estelle to follow her instinct, though, was having good results. This energy, a providential and life-preserving force, is more powerful than medicine and physicians. After taking all necessary precautions to ensure a safe arrival, I was confident that, if the need arose, Estelle could take care of herself on the road. She herself felt safe and seemed unconcerned about dangers of the trip. We had discussed potential problems, but she had dismissed them. Therefore, Madame L'Hardy and I decided that her daughter would ride behind me on my stable, reliable horse.[89] Madame L'Hardy would have her own mount. To guard against possible mishaps, we planned for the horses to maintain a slow pace. If it were necessary, I could strap Estelle to me by a large belt and thus be aware of all her movements.

While in the crisis state, Estelle was seated on my coat behind me with her feet in two small stirrups. Because Estelle never liked being tied up, I attached my handkerchief to my belt, and she held it with one or both hands. The journey began at a slow pace, but it was going so well that, when we reached a stretch of meadow, I encouraged the horse to trot and then gallop in order to keep pace with Estelle's requests to speed up. The quicker pace did not lessen her courage or throw her off balance. Indeed, she got down from the horse several times to gather flowers and remounted without experiencing the slightest fear. In fact, Estelle acted like a fine horsewoman during this one-day trip of almost 30 kilometers. It was as if she had known how to ride a horse her whole life. Consequently, our trip to Annecy was not tiring. Moreover, when Estelle awoke from her crisis, she found herself in a new, completely unfamiliar place. She had forgotten everything that had happened since the moment she went to bed the night before—the goodbyes in Bauges and all she had seen and done en route.

The few days we spent in Annecy gave her much happiness, such as the feast of Corpus Christi, a holy day that seemed to interest Estelle. The weather was wonderful, and we ran errands in the carriage along the shore of its beautiful lake. Her only regret was that she might forget most everything she saw, heard, and learned during her somnambulistic state. Estelle said, "If this continues much longer, how will I finish my education? And all my childhood friends will have learned more than I have!" This concerned her greatly, and with good reason. For, during the 21 days since we had left Aix until the moment we returned from Annecy, Estelle didn't have the slightest memory of what had happened to her and around her while she had been in crisis.

Upon returning to Aix on May 26, Estelle resumed her treatment, including a daily swim in the pool for one to three hours. At first, she needed our safety equipment to hold herself up in the water. Shortly thereafter, Estelle no longer needed the safety gear while in crisis. She became one of our best swimmers and enjoyed giving lessons to the young people in the baths. Estelle participated in activities typically done in the water—diving, swimming, and floating. When she wanted to dive, she prepared herself by closing her eyes with a self-devised magnetic formula (see my endnote 11). Estelle's eyes remained open during somnambulism; she rarely blinked and only to wet the eyes. Since she feared getting water in her eyes, Estelle instinctively closed them in order to dive. By using another magnetic formula to undo the first one, she opened her eyes again after the dive and continued swimming.

Every day thereafter until June 30 when Estelle returned to Neuchâtel, the integration of the crisis and non-crisis states continued incrementally. The highly elaborate electrogalvanic phenomena Estelle had manifested gradually lessened in power and intensity; eventually, almost all of them disappeared. Thus, by May 29, Estelle no longer suffered when she tied black shoe ribbons made of silk around her cotton stockings. Also, between June 10 and 21, her crises would stop while she swam in the pool because of the decrease in the electrogalvanic phenomena. Therefore, in order not to drown and to swim more easily, she was forced to reenter a crisis state. Beginning on June 19, silk, that would "burn" her if it touched her skin, no longer stiffened her limbs, and on June 21 she swam without a crisis.

Nevertheless, on occasion during her last months in Aix, Estelle manifested some violent phenomena, albeit polarized. On June 17, she had another attack of rage caused, like the preceding ones, by a domestic dispute that deeply wounded her pride and pushed the limit of her extreme vulnerability. Meanwhile, Estelle was tormented by the desire to magnetize another cataleptic[90] to calm her crises and anxiety. This patient had been referred to me by Professor Breton, physician and *Recteur de L'Académie de Grenoble.* Estelle got along very well with this patient and wanted to spend the whole day with her. We did not let Estelle magnetize her, however, since the influence of one child on the other (whether positive or negative for Estelle), would deplete Estelle because she would need to mobilize energies and passions (see my endnote 12).

To summarize Phase Two, I make three observations concerning Estelle's general physiological phenomena, some of her remarkable physiological phenomena, and the general progression of her illness.

General Physiological Phenomena:

- Somnambulism achieved at will;
- Deepening or departure from somnambulistic state achieved at will;
- Phenomenon of magnetic attraction and repulsion;
- Transfer of hearing[g] to different parts of the body; particularly to the wrist, elbow, or epigastrium[g] during the active crisis but always to the epigastrium[g] during syncope[g] or ecstasy;
- Dietary shifts: diet entirely different in the crisis state and the waking state; one state (including what Estelle consumed during it) does not impact on the other, even though the shift from one state to another is instantaneous;
- Momentary (and only accidental) alteration of the functions of the mind;
- Hallucinations, rage, and fury;
- Irresistible need to run;

- Extraordinary development of the functions of the mind;
- Her narration of her own curious medical history, from its onset;
- Phenomena of imitation and echo;
- Four states of the illness, each distinct from the other three:
  - the state of active somnambulism (or active crisis without suffering),
  - the state of inactive or dead somnambulism (lethargy or syncope[g]),
  - the state of tetanic-spasmodic crisis (with suffering),
  - the state of cataleptic crisis (without suffering);
- Sensation of a ball bursting in the esophagus which always brings about a noticeable improvement;
- Fatigue when she hears the screech of a knife cutting hard bread;
- Convulsive laugh;
- St. Guy's Dance[g] by imitation;
- Increased skill in swimming and in riding horseback while in crisis;
- Calming effect of music, silence, darkness, and solitude on most emotional rages, a reminder of the therapeutic effect of David's harp (tarentism[g]) on King Saul;
- Disappearance of fear and panic during crises.

Some Remarkable Physiological Phenomena:

- Electricity. Touching silk, a cat, fur, a long-haired dog or a water spaniel "burns" her skin, stiffens her, and puts her into a crisis. This predisposition disappears as the treatment progresses;
  - Glass "burns" her, porcelain seems warm to her but doesn't "burn"; cups and bowls made of pipe clay and pottery seem cold to her;
  - All crystallized objects tire her, although a mixed crystallization bothers her a little, and a perfect crystallization bothers her the most. Thus, a rock crystal feels extremely heavy and "burns" her, whereas ground silica may seem somewhat heavy without "burning" her;
  - Chalk does not harm her, but crystallized calcium carbonate increases her anxiety without tiring her as much as silica crystals;
  - The mere presence of the colors red and violet tire her;
- Galvanism. Pure gold is a true fulcrum for Estelle—with it, she could lift the world. "A quintal[g] of gold," she told me one day, "would give me enough strength, I believe, to lift up the Church in Aix. This little crystal" (which weighed almost nothing[91]), "seems

heavier to me than all the gold I am covered with."[92] The juxtaposition or soldering together of two metals produces in Estelle very subtle electrogalvanic phenomena. Hence, while in crisis, she refuses the gold-coated silver teaspoons that she had found agreeable during the waking state. It is undoubtedly also because of these subtle electrogalvanic phenomena that we notice the differential impact on the patient of a stopped watch, a watch that works well, and a watch that works badly;

- Animal Magnetism. Although Estelle can now enter a crisis state on her own (fascination takes her there against her will each time she has an opportunity), she still easily responds to magnetic passes to enter that state.

General Progression of the Illness Estelle's state of exaltation—the worst of the problem—continues until mid-April. At this point her symptoms decrease, and gold is less necessary to Estelle. From this moment on, the crisis state begins to blend continuously with the waking state so that the various manifestations of the crisis state increasingly resemble the normal state. Everything moves toward bringing the patient back to her former health as the blending continues.

Phase Three of the Magnetic Cure after Estelle Left Aix-en-Savoie: June 30, 1837 through April 1838

Phase Three includes phenomena after Estelle's departure from Aix on June 30, 1837[93] until the end of April, 1838, when the patient was preparing to return to Aix to complete the treatment.

Within Estelle, during this ten-month interval, the crisis state almost completely integrated with the waking state. In other words, the waking state with paralysis of her legs incrementally fused with the state of exaltation in the crisis state. Little by little, as an overstimulated state released its energy to the under-stimulated one, Estelle gradually returned to an almost normal state, although she retained some aftereffects of her long illness. In Phase Three my commentaries are brief since after June 30, 1837, I made no direct observations. The summary of our little patient's progress will thus be completely derived from my correspondence with her respectable family. I report sections of these letters verbatim. I do so to express, as accurately as possible, the feelings of Estelle and of those around her, particularly regarding some new phenomena that emerged. These converging reports also provide further proof of my assertions.

In parallel to my work, those in the medical establishment in Paris[94] and allied learned societies are currently debating matters under discussion here: galvanic and electromagnetic phenomena; spontaneous and induced somnambulism; and the transposition of the senses.[g] This timely monograph on Estelle will interest these professionals. It can promote understanding among those who are polarized and help to bring them together. Only when we understand each other well can we get along. Then, we can realistically trust that science is steadily moving toward Truth, our anchor of safety.

Estelle left Aix with her Mother on June 30,1837. They had a good trip, in short and easy stages, and arrived in Neuchâtel on July 2. Estelle delighted her family who, although certainly expecting her to have improved, could not conceive that she had made such extraordinary progress. Her Aunt Julie must have been the most joyful, for when she had left her niece in December, Estelle was a cripple. Furthermore, her aunt had been unhappy about the extended stay in Aix, convinced that our efforts would be in vain. She and the entire family believed that Estelle suffered from a profound alteration in the lining of the spinal cord. Aunt Julie had believed that prolonging the time in Aix was very likely a waste. She had agreed to the stay only out of affection for her beloved sister lest this devoted Mother despair. Julie believed that this good, concerned Mother was blinded by the false hopes I promoted. In reality, I had encouraged Madame L'Hardy to consider Estelle's recovery as only a possibility (see my endnote 13).

On July 4, spontaneous somnambulism completely stopped. Estelle retained, however, the ability to put herself into the crisis state voluntarily. She developed a cough that at first seemed phlegmy and catarrhal,[g] but her cough disappeared after a cold swim in Lake Neuchâtel on July 11.[95] Meanwhile, Estelle's gait was still unstable. She was not interested in her studies. Daily, she played age-appropriate games and enjoyed herself with her cousins. Nevertheless, she put herself into crisis every evening to amuse her Mother with jumps and skips; she amused herself by proving to her Mother that she lost nothing by letting her strength lie dormant during the waking state. Not for anything in the world would our little mischievous child have wanted her family to think that she enjoyed more muscular strength at one time than another.[96] Estelle was troubled because she sensed she was being spied on, and she felt that she was not believed. As a result, she wanted to repay someone in kind. Later in the story, when the tables were turned, Estelle put herself into a crisis for the entire day to prove she had not lost her strength, even though this loss would have had a legitimate cause at the time. But those who mistrusted Estelle's account of her recovery from paralysis annoyed her. She persevered without ever giving up and reserved all her somnambulistic antics for her Mother.

On July 21, 1837 our dear patient fell on her back in a way similar to her fall at the genesis of her paralysis on November 27, 1834. This recent fall was with the same friend! These converging circumstances made a painful emotional impression on Estelle. Within two days, however, the pain had almost completely disappeared without any negative aftereffect, either emotional or physical. Nevertheless, the smallest pressure on her back still caused brief crises of catalepsy. During those somnabulistic crises, her young cousin (Mademoiselle Marie B.) was the only person who could approach her, relate to her, or talk with her. The sight of red fabric evoked in Estelle the same nervous symptoms as any pressure along her back. She was repulsed by meat or broth. Seasonal fruits, eggs, milk, salad, and bread continued to be the foundation of her diet.

On July 26, we climbed the mountain of Tourne, situated on an elevated plain in the Jura mountain range that dominates the Neuchâtel Basin. Each day Estelle and her young friends gathered flowers, picked and ate strawberries, raspberries, and blueberries, and drank cold milk and cream. They would walk almost from dawn to dusk, run around, and take rides on horses or mules. After July 29, Estelle no longer put herself into a crisis state to amuse her Mother. But, when she became too tired in the midst of her mountain outings to return to the lodgings, she resorted to putting herself in crisis so that the physical exertion would not tire her.

In the month of August, during the *alpage* season, Estelle remained in the mountains.[97] There, Estelle usually awoke between nine and ten o'clock. Formerly, she had been extremely afraid of cows and bulls but cured herself of her phobia of these animals—astonishing her entire family. Estelle's physical stamina improved considerably with the Alpine exercise to which she gave herself in complete freedom.

The weather, that until then had been nice, deteriorated. Estelle was particularly drained by the stormy weather on August 3, 5, 8, 9, 11, and 13–15. The days in between, which were fairly nice and serene, put her back into her normal state. During the storms, however, her state of crisis, that renewed itself with each lightning bolt, showed such remarkable symptoms that I feel compelled to describe them. I choose the stormy days of August 13 through 15 because they illustrate the electrical phenomena Estelle exhibited in Tourne during this first half of August. I use the words of her Mother who sent me this account from Peseux dated August 18, a few days after this incident:

> As soon as the weather got stormy, my daughter immediately fell into catalepsy. Her crises lasted for 20 to 30 minutes but stopped as soon as the storm struck. Sometimes we found ourselves in the midst of thunder clouds. Thick dark clouds opened now and then; we could see lightning fly all around us. When the storm was at its peak and everything seemed to whirl and crash around us, Estelle, in the crisis state, would stare at the lightning. She would startle, making convulsive movements with her hands and arms, falling to the ground, tearing out the grass and clinging to it, crying out in pain. She would do magnetic passes on herself as if to lift an oppressive weight. She neither could nor would have contact with me. Her eyes were fixed and immobile, as when she is the ecstatic state. Yet, at the same time, her body was not tense and could move in all directions. After the 45-minute thunderstorm, Estelle emerged from her violent state with no memory of the experience. She felt only fatigue and achiness in her entire body.[98]

During these crises provoked by the storm on August 15, Estelle exhibited some unusual symptoms. The crises were no longer standard attacks of catalepsy but rather a derivative cataleptic state in which her body remained supple, her eyes open and staring. She heard nothing. She also appeared to be unaware of what was going on around her. Nevertheless, she pushed away anyone who tried to approach her. If the person persisted, Estelle seemed to suffer even more, for her face took on a pained expression. At those times, she would place on her head cold water compresses that seemed to comfort her. When it would hail, she would ask for some of it, eat it avidly, and fill a cloth with it. She would then put the ice pack on her head. At that point, even though the storm was still raging, she would immediately recapture her gaiety and come out of the crisis state. She would have no more crises the rest of that day.

On August 20, there were no crises. After nightfall, however, Estelle manifested a new phenomenon remarkable in the story of somnambulism: Suddenly Estelle calls me [this is the verbatim rendition of what Madame L'Hardy wrote me about this phenomenon] "Mommy," she said to me, "this is very strange!——I am not in crisis, yet I see you with my eyes closed——I distinctly see everything around me——I couldn't read, but——I see——with my eyes closed——without being in crisis; that is for sure!!!" In order to assure myself of this and to be able to recognize the truth of an incident I myself did not believe, I led my daughter to the darkest place in the apartment. There, I opened a book at random, and without hesitation Estelle counted the number of lines in a given space. I experimented with this several times. Each time, Estelle gave the correct number of lines.

An hour later she came to bed. She went into a crisis to amuse me a bit and to lessen the anguish I had felt since the beginning of the month. Oh, how we missed you at that moment, Monsieur Despine!!! Estelle always pushed me away during the crisis state. I could not be in contact with her during this long drama, and I can give you no information about what was going on within this little being—this frail, delicate, and very sensitive child. We were forced to let her be. While in Aix, Sir, I was not anxious, no matter what happened. Here, however, I had to be brave in the face of adversity. But, patience. The stormy season has passed. I hope that no more storms will come and that Estelle will be calm from now on, at least as far as the weather is concerned.

We had hoped, dear Sir, to be able to live in the mountains in perfect isolation, completely ignored, but such was not to be. People notice us here—despite all our efforts to remain anonymous. On this mountain, several people stopped us on our walks to ask, "Is this young lady not Dr. Despine's somnambulist from Aix?" People said a thousand and one things about her that neither you nor I would know about, Sir. They summarized all the very extraordinary things somnambulists past, present, and future have said, do say, and may one day say and do. All this, as you can imagine, utterly exhausts me.

On August 31, we came back to Peseux. In Tourne, we had no cats; here, however, they abound in order to defend us against rodents. The cats come and go freely in Estelle's room. They go under her chair without causing her suffering, although she refuses to touch them. The sensitivity on her back has greatly diminished. I often touch her there without her noticing. The sensitivity to touch on the chest remains.

Both night and morning, Estelle avidly drinks a big cup of hot milk from an animal's udder. She still rejects meat and broth, but since she is doing so well in other ways, we go along with her preferences. For the last few days, though, she has developed a taste for ham, and eats it with great pleasure. It is the only meat she has had since the onset of this long, cruel illness.

Estelle was happy, Sir, to find several of your letters when she arrived here in Peseux. She reads and rereads them. Doing so invigorates her. Moreover, every night without fail, both here and in the mountains, she wishes you a good night before going off to sleep. She also sends you a goodbye kiss, straight from her grateful heart.

Toward the middle of September 1837, Madame L'Hardy wrote:

We went to see the De Meuron ladies. We were happy to talk about all we owe to our stay in Aix. We also visited Dr. De Castella, who literally could not stop looking at Estelle, admiring her strength and her good appearance, since both have improved considerably this summer. For the last few days, however, she suffers because of her teeth. Monsieur De Castella advised her to return to Aix, but her little somnambulistic pride was hurt, and she retorted, "Of course, I will return to Aix, but I will know when it is the right moment for that."

Since the problems with her teeth began, she has had no more crises. Estelle still has the ability to put herself into the somnambulistic state at will, but I must be alone with her. Under no circumstance does she want anyone else to see her with more strength at some times than at other times.

At the end of September, again Madame L'Hardy wrote me:

We thought, Sir, of spending the end of the autumn in Aix, as we had planned when we departed last June. Since that time, however, Estelle's health has continued to improve daily. So we think that a stay in Aix would be useless right now and think that it would be wiser to wait until spring. Even Estelle favors the spring. If there were a noticeable alteration in the health of my daughter, I am sure that my whole family would insist on my bringing her quickly back to you, Sir, as you are the only person in the world capable of healing her, with God's help. The present strength of your patient proves this point again and again. She has continued to improve even since my last letter, although I say nothing about it to her. She wishes that things continue to go badly.[99] We weighed her upon our return from the mountains and found that her weight had increased by four Swiss-Neuchâtel pounds[100] since she left Aix and arrived in Switzerland. Meanwhile, Estelle insists that she has lost much weight. She sleeps well for ten hours at a stretch. She has an appetite, but still is repulsed by meat. She is outdoors for almost the entire day without ever complaining of fatigue. She has grown, and her stride is more confident and strong.

Crises, or I should say, little crises, are rare. Estelle is still able to self-induce a crisis state when she wants to. She enters one spontaneously almost every day. But the difference between her strength in crisis compared to that in the waking state is hardly noticeable. Despite my experience, I am unsure whether or not she is in crisis unless I see her going upstairs. While not in the crisis state, Estelle is able to go up and down from my room to the garden several times every day, despite the more than 60 steps. Thus, you see, Sir, that I have many reasons to remember you and to bless you and my fortunate stay in Aix. I especially want to complete the treatment that astonishes and confounds all those who have known the details of this long, dreadful illness. The facts speak so clearly here that the few remaining nonbelievers are quickly silenced.

In a challenge to magnetizers, somnambulism, and the so-called transposition of the senses,[g] Dr. Burdin of the *Académie Royale de*

Médecine de Paris offered a prize.[101] He would award 3,000 francs to anyone who could read without the help of eyes, touch, or light. I quickly informed Madame L'Hardy of the contest, because no one seemed to deserve the prize more than Estelle. This excellent Mother responded in a letter dated October 17, 1837:

> I received, Sir, your letters dated October 8, 9, and 10, and I must begin by thanking you for the details included and for the interest you continue to show in your little invalid. I am aware of how deeply indebted I am to you, Sir. Without your devoted efforts, far beyond those one would expect from the most zealous of doctors, I would not have the happiness of seeing my Estelle so healthy, enjoying her life and her family against all odds. My dear Dr. Despine, the deep gratitude in my heart makes it painful for me to deny you what you ask. The more I think about Dr. Burdin's challenge, however, the more the idea of Estelle's being on stage repulses me. If she were a little boy, it is likely that I would think differently. But a girl must try to keep a low profile and avoid being spoken of, particularly when the events might develop her pride. I fear, with reason, that such a public demonstration might give her an exaggerated sense of self-importance. Moreover, if this exposure did not affect her attitude, it might, at the very least, compromise her future.
>
> I assure you, Sir, that I need very strong reasons such as these to refuse something that seems so very dear to you. In this demonstration you see something for the well being of humanity. Ah! my dear Dr. Despine, if you knew how much it pains me to refuse you!!! You would surely have pity on me if you knew!!!
>
> Estelle continues to get better. That she dresses herself with extraordinary care makes me think that the sleeping and waking states are increasingly integrating. Often when she is uncomfortable, she splashes herself under the spigot of the pump. This practice always does her a great deal of good.

At the end of November 1837, Estelle's Mother wrote:

> Sir, I have delayed contacting you, but I have lost several relatives and family friends, one after the other!! In the midst of these scenes of mourning, I have often thanked God for sending me to Aix. If He had not led us to you, Sir, it is very probable that Estelle would no longer be alive!!! Now I only rejoice about her good health. All this is to say that the winter has not yet really challenged her. Indeed, her strength has not decreased at all. Quite to the contrary—it seems to be following a healthy course, though perhaps somewhat slowly, for she suffers from the cold and complains about it a lot.

It is unfortunate for Estelle that she does not want to go into the crisis state. Doing it repulses her so much, however, that I do not insist. I let her do as she wishes, particularly since I see her continuing to get stronger and returning to her healthy routine. She once again enjoys craft projects, and she works on them even more than I like. She is not yet concerned about lessons meant to cultivate her spirit, nourish her heart, and develop her intelligence. I am looking forward to her interest in those returning.

With great enjoyment, she watched the snow arrive. Snow continues to please her, but only as something to eat and to cool her milk. Yet, she no longer rubs herself with it. In addition, Estelle no longer pours cold water over herself or takes cold showers. I let her decide about these matters, but I have some concerns because blood often rushes to her face. To remedy these rushes, however, I use cold water and vinegar to wet the thick compress she requests. I do not squeeze it out. We refresh the compress each morning after she gets up.

The cataleptic crises, as well as of those of somnambulism, become increasingly rare. You see, therefore, Sir, that all in all Estelle's health is in the most satisfactory condition. It is with the most profound and sincere gratitude that I tell you that her well-being surpasses all my greatest hopes. Dr. De Castella would agree. We have just seen him for the fourth time since our return. He was rightly struck by the strength of your "Little Resurrected One." Perhaps he thought, like so many others, that Estelle would weaken once she was away from the magnetic influence. Facts prove the contrary, however. It is impossible for someone to remain doubtful when contrasting my daughter's state before her departure for Aix with her state upon her return.

Peseux, January 3, 1838

Sir, I do not want any more days to pass without extending our family's New Year's wishes. May God grant you a good and happy New Year. May He also free you from the worry and pain that have often and cruelly beset you these recent years!![102]

No doubt, Estelle will be writing you to tell you herself about the joy she felt the day before yesterday, thinking that she can now say to you, "Yes, I will come to Aix this year!" She did not experience the arrival of young Sophie La Roche at your home without some jealousy. This feeling, however, quickly passed. She is getting used to this idea and understands that she cannot be the only one to profit from your devotion to human suffering. She has even begun to like your young patient, and she greatly desires her healing. This wish is perhaps less for Sophie than for you—she wants you to triumph over those who do not want to believe either in the instinct of self-preservation or in the instinctive impulses of somnambulists and cataleptics.

Photo 10 Drawing of patient Sophie La Roche from Virieu. Artist unknown.
Source: Courtesy *Archives des Thermes nationaux d'Aix-les-Bains.*

I failed to locate the newspapers you mentioned in your last letter. It is with those newspapers as it is with the famous one that printed a story about Estelle without consulting me or you. Many people read the articles but no one knows how to get hold of them.[103]

I thought that you were hoping to rest this winter. But, you have new subjects to study. I doubly admire the devotion and zeal you always put into making your life useful to mankind and to advancing science. Also, Sir, I send all my good wishes for Sophie's cure, hoping with all my heart that you will have the same success as with Estelle.

My daughter continues to do admirably well, even though the winter is trying for her. It does not decrease her strength, although it prevents her from leaving the house. She also eats little and hasn't grown anymore. Yet, from one month to the next, I observe a noteworthy increase in strength. Crises no longer occur except by accident, such as when she is hit on her back or chest or when she has contact with fur. Since we avoid such incidents as much as possible, the crises have become extremely rare.

For the last few days, her chest and back have been covered with big red stretch marks similar to those of 8 or 10 months ago in Aix. I am not worried, however, since Estelle is experiencing no pain and since she sleeps normally. I am viewing this eruption as a salutary effect of Nature. Estelle's diet is still the same—salad, cold milk, beer, and eggs—and meat and broth continue to disgust her.

So you see, Sir, that Estelle is having a very good winter, all things considered. I hope that a few more weeks of your care will secure her health with the completion of her treatment. Ah! Sir, how I thought of you on December 31, Saint Sylvester's day. What a happy anniversary with good memories. I felt the need to sanctify the day, so I went to Montmirail,[104] where I was able to do ritual observances of piety and edification to commemorate the day. We welcomed the New Year, celebrating the glory of our most holy God, whose works are so admirable and who deserves the expression of my deepest gratitude.

February 12, 1838

We have just had a bout with the flu that passed through the entire household. I did not want to tell you about this, Sir, when you are so busy writing Estelle's story. I did not want you to wonder about the certainty of her cure; I have not had to doubt it despite the fact that this unexpected catarrh[g] greatly wearied Estelle. In fact, hers was one of the more serious cases of the flu. But Estelle tolerated it perhaps better than her little cousins who had contracted it before she did. Of course we treated her without a doctor. After you, Sir, she refused to see any of them, even the excellent Dr. De Castella. Thank God, her instinct (of self- preservation) guided her perfectly, even though she did not resort to making use of the crisis state during this entire period. In fact, as soon as the catarrh[g] appeared, Estelle no longer wanted anything but hot drinks; she renounced beer. But, as soon as the fever was gone and fits of a purely catarrhal cough had replaced those of her nervous cough, Estelle cried out vehemently for snow. It did her a great deal of good. Consequently, the catarrh itself has become for me the most unequivocal proof of a complete cure.

I will tell you also, Doctor, that for me another proof of her cure is Estelle's return to her former docile, obedient personality. She now lets me be her Mother, and she tolerates all my observations and advice. May it be God's will that you have as much success with Sophie La

Roche. Her healing would be a great accomplishment for humanity to whom, without exaggeration, you have dedicated your life and sacrificed your leisure. Nevertheless, you have still not triumphed over the disbelief of most of your colleagues. If only Sophie, who has been crippled for more than four years and who has frustrated the skill of all physicians who have treated her for the last eight years, were restored to health. I believe that Sophie and Estelle will prove to your colleagues that sometimes it is good to deviate from the usual path—and that with patience, perseverance, and a good method, one ends up triumphing over even the most serious illnesses.

April 18, 1838

I received, Sir, your letter dated March 25, along with your first two pages of Estelle's story. I read this account with great interest. Moreover, in going over all the stages and phases of this cruel illness I am very aware of all that I owe to God and to your care. Estelle and I will never forget what you have done!!!

I like to hope that Estelle's example will become useful to humanity by enlightening those in the medical establishment about a type of paralysis unknown to date. I also hope that many parents will avoid the cruel anguish that was my fate before following the treatment methods you prescribed. I sincerely rejoice about it, Sir, for you and for them. That Estelle has begun requesting lessons is proof of her good state of health. Moreover, the facts speak authoritatively here to all who had condemned her to a premature death or never to walk again. Estelle's good health proves the power of your methods—not those formerly used to heal paralysis. I wish strongly, Sir, for Estelle's future, that all the extraordinary events that happened to her remain unknown in our canton. On the other hand, where she is not known, I will certify to all wishing to know that all events you recount (however incredible and marvelous they may seem) are the purest, simplest expression of what happened during this singular illness.

I am longing for you to see Estelle again, Sir. She is taller, stronger, and well. You will enjoy, I hope, the success of your fine care. The inhabitants of Aix will be able to believe the truth of this healing, even those who said Estelle pretended to be sick and then cured by "Good Father Despine," who was the "dupe of a child's cunning." My daughter, in fact, is feeling as well as could be expected. Her improved condition will, without a doubt, give hope to your other patients whom we long to know. It will also be a source of embarrassment for those who had predicted that a cure of this kind could not last long.

We intend, Sir, to leave by coach for Aix on May 1, and arrive in Geneva on the same day. The next day, God willing, we will be with our "Good Father Despine." This trip, as you can see, will have nothing in common with the one we made in July, 1836 with her mysterious basket. You cannot imagine Estelle's joy simply at the thought of seeing you again. I also long for you to see her so that you may rejoice in your work. She continues to get stronger despite the 21 days of flu when she would drink only hot milk. Obviously, this flu greatly weakened her. She made up for this temporary lack of strength by putting herself in a crisis state for the good part of the day. Consequently, despite this long, severe abstinence from solid food, all of us at home found her to be as strong and as alert as she had been in her usual state before the flu. It should come as no surprise to you that she was walking very well, but tired quickly. Nevertheless, Sir, she gets stronger every day and retains no nervous symptoms except an ability to put herself into a crisis state at will. Since January, I have not seen the slightest evidence of catalepsy. This, Sir, gives me great hope for all your patients.[105] Their cure particularly interests me because of the pain I endured and all you have done for my poor Estelle.

In the hope of seeing you soon, Sir, and filled with the most heartfelt gratitude I have the honor of being,

Madame L'Hardy

Concluding Reflections

I will now end this long, curious story with some reflections inspired by the study of the singular electromagnetic phenomena that have I observed. They somehow seem to belong to this pathological state of the nervous system. As I stated, the illness I describe is undoubtedly not new. I do not believe, though, that any modern or ancient author has ever described it. Moreover, the kind of paralysis Estelle L'Hardy had is definitely not due to compression of the brain or the peripheral nerves. None of the therapeutic methods recommended for such cases worked for her. Estelle's type of paralysis belonged, therefore, to another order of pathological phenomena, although an order difficult to identify. Estelle's neuropathology seemed to be entirely the result of the poor circulation and maldistribution of nervous fluid throughout her body. My conclusion is supported when we look at the influence of animal magnetism on our young patient and when we consider the action on her of electric current, cold water, ice, and snow. It would appear that the simultaneous use of the springs, magnetism, electricity, and the application of cold—all redistributed the nervous fluid according to Estelle's instinct of self-preservation. This instinct is developed to its greatest degree in the curious state of somnambulism.

Most of these events are unknown to physicians who bother only with theory. They have acquired the unfortunate habit of wanting to bind Nature to their own methodological framework. The intention of these theoreticians to facilitate the study of science actually limits it. Unfortunately, these same events are also little known to most physicians in clinical practice who have many patients. Either these clinicians have not had such patients in their practice often, or they may not have had sufficient time to attend to them. Also, some physicians may be satisfied to marvel at phenomena rather than investigate them thoroughly. They consider medicine an industry rather than a science.

As for me, my dear reader, who has sought to find only the Truth and to present it to you in all its glory, I have God as my witness that many times I have observed certain unusual phenomena in Estelle. I had already more or less recognized these same symptoms in many other patients suffering from analogous ailments (see my endnote 15). These phenomena—whose constancy, description, and course are remarkable—denote a pathological state in its own right that deserves to be studied with care. Illnesses of the nervous system seem to be increasingly common in all social classes. Moreover, these electromagnetic phenomena deserve more complete consideration by physicians, philosophers, and physicists. It is a new world, one that the author of Nature leaves to our investigation and research.

The external parameters of this new world are already known: its pitfalls, its inaccessibility, its mountains, and their summits. This new world, however, has been observed from afar. It has been viewed from the perspective of established and imperfect instrumentation that directs and limits our observations. The internal parameters of this new world have not yet been successfully explored since our methodological compass and guides are flawed and lacking. These guidelines need to be developed to pave the way for the physicians of the future. We are at the dawn of an era when new, more trustworthy methods will be at our disposal. This cumulative knowledge, developing over more than half a century, results from the serious investigation of phrenology, galvanism, electricity, and magnetism, possibly a modification of magnetic somnambulism. This progress also derives from the analytical and research-based spirit of today's students. This spirit, if directed and strengthened by good philosophical studies, should bring about more complete, detailed findings. Only commitment to study can appropriately harness the love of novelty that makes young people run frequently after chimeras, seeking avidly the glimmer of glory rather than the benefits of a solid reputation based on an arduous education. An unbridled fascination with novelty encourages a frivolous spirit that explores nothing in any meaningful depth.

Photo 11 Antoine Despine at around 73 years old. The photograph is believed to have been taken by Constant in Antoine's medical office at the Baths.

Source: From the private archives of Philippe Despine, Dijon.

If only the coming generation would bring to its studies the spirit of wise observation and to its research a discerning, critical spirit. If only the new generation would not dismiss experience by relying on vain theories only to simplify or shorten a work. In medicine, more than in any other science, experience alone can serve as a foundation for all doctrine and systems of education. If this generation has the wisdom to exploit experience and observation, great discoveries await the new medical world. On the other hand, the new generation may be apathetic and disinterested. It may scorn the writings and observations of its elders, or neglect what its venerated predecessors have learned through concentrated work. The new generation may want to know everything without ever delving into a topic, using superficial, encyclopedic analysis or books, rather than studying Nature—the true source of all knowledge. Finally, the new generation may be satisfied to judge superficially according to what others say. They may not want to be trailblazers whose personal experience guides them. If all this is so, despite their advantages, the new thinkers will not know how to navigate the stormy sea of life. They will risk peril or lose track of their objectives as they explore the sciences—and their exploration will be filled, as always, with pitfalls, setbacks, and illusions.

Despine's Endnotes 1 and 2 have been integrated into A Curious Case of Neuropathy

ENDNOTE 3

ELECTRIC AND GALVANIC ACTION OF METALS

Estelle presented numerous relatively unusual nervous phenomena: pure electric, galvanic or magnetic, and mixed phenomena of electrogalvanism, electromagnetism, and others. All these, occurring successively or simultaneously, are themselves very curious. To do justice to each, I could fill huge volumes with anomalies and variations of nervous phenomena that could emerge in succession, either suddenly or gradually. I will not, however, undertake that task; it would take me too far off topic. I will limit myself, therefore, to the most interesting new phenomena and those rare in form and presentation. I will discuss phenomena related to physiology and, in particular, phenomena apparently resulting from what I call physical and mechanical action on body reactivity (I omit phenomena related to pure pathology). I observed how the action of metals (with their electric or their galvanic forces) and mineral substances (with their natural structure and degree of perfect crystallization) act upon the body.

In nervous ailments, it is difficult to draw a clear line between phenomena belonging to physiology and those belonging to pathology (the sickly state). Estelle had been quite ill for several years; consequently, everything about her was abnormal. For Estelle's story, however, I will classify every day phenomena under physiological phenomena. They presented themselves without real organic alteration, despite the patient's extraordinary state of exaltation. But, this physiological category presents several functional aberrations that diverge from the ordinary, natural order of the healthy state and are true pathological phenomena. I do not intend, however, to debate this point; understanding it will suffice.

I wrote that our warm baths and showers never caused heavy perspiration in our young patient from Neuchâtel (as is usual for other bathers). Nevertheless, she could tolerate the waters only between 28° and 30° C. When electrification went beyond a specific intensity, it provoked a state of lipothimy[g] or syncope,[g] analogous to the first degree of catalepsy or somnambulism. I also noted that Estelle ate nothing all morning and that a lean diet suited her infinitely more than meat, fatty broth, and similar foods—all these tired her terribly.

I also wrote that, until December 22, 1836, Estelle suffered greatly each time we tried to move her. She did not allow anyone to touch her back. Dressing, undressing, and settling into a lounge chair were a real martyrdom for this young person. She had been unable to move in bed because of the application of moxas,[g] and since her arrival in Aix she would faint in any position but a horizontal one. The sensitivity of her back was so intense that she allowed me to touch it only once during her first five months under my care. She let me touch her back only to escape additional moxas[g] like those applied in Neuchâtel. I assured her I would not use them unless all else failed. True, we attributed to exaggerated sensitivity Estelle's consistent refusal to cooperate with our examination. What followed, however, proved to us that what she told us was perfectly true. She suffered because the skin over her entire back was profoundly sensitive. There was no profound affliction of the skeletal system, as had been thought.

As soon as the vital electrogalvanic action occurred under magnetic influence, the hypersensitivity of the skin disappeared. Our young patient allowed herself to be touched on all points of the dorsal area from the back of the head to the sacrum. She reluctantly allowed a complete physical examination of her vertebral column and thoracic chest. The freedom to touch her, however, stopped when the magnetic influence stopped and resumed when she was again magnetized. Estelle was still so sensitive on December 22 that we had to wrap her completely in down and cotton-wool padding and surround her with jars of hot water. Otherwise, she got cold.

I described the first magnetic session that occurred on December 22 (see Treatment, Last Stage). The young patient laughed at the process until the twentieth minute. She then began to feel the action of this new treatment and to appreciate its value. She made a complete turn-around in the way she spoke of it. She fell into a light sleep. From this session on, the first in a long series of electrogalvanic-magnetic phenomena appeared. Estelle had been in Aix since July 15. During these five long months, she had gained little from her treatment. Her limited progress amounted to nothing compared to the developments during the last days of December.

On December 23, Estelle's sensitivity to electricity increased considerably, and magnetism already elicited remarkable phenomena in her. Sleep was almost complete. She clearly saw and recognized the fluid leaving the magnetizer's fingertips. Since she no longer needed the down quilt and hot water bottle while under the magnetic influence, Estelle had them taken away. If Madame L'Hardy came near her daughter, the mother's "atmosphere" tired Estelle. Estelle began to prescribe what she needed. She affirmed that the lean diet suited her state, and she saw—eyes closed—what was going on around her. She indicated the kinds of magnetic passes to be done on her. She felt the magnetism gradually act on her and felt the progress it made. Even though Estelle had no medical knowledge, her recommendations were medically sound. Nevertheless, she felt that the process could happen only gradually and that we must not try to accomplish everything at once. She felt that certain passes and kinds of magnetic pressure have varying effects on her. Finally, Estelle described very accurately the magnetic effects that her magnetizer's warm and cold breath had on her.

On December 24, Estelle had her third magnetic session. She had enjoyed a wonderful night's sleep the night before. She felt that electrification was becoming less necessary as magnetic action had more effect on her. She fell asleep that day at the second minute. At the sixth, her eyes were "nailed shut". She perceived grains of fire. Near the tenth minute, convulsive and involuntary jumps were noticed, especially in the paralyzed limbs where the magnetic passes were being done. At the fifteenth minute, she saw everything going on in the room, although her eyes were shut. At the thirty-fifth minute she guessed the thoughts of someone present, and at the fortieth, she had a succession of fantastical visions of all sorts. At 60 minutes, she perceived luminous points and electric aigrettes[g] at the magnetizer's fingertips, still with her eyes "nailed shut." Fifteen minutes later Angeline, her protecting spirit, appeared to her. This familiar spirit, as Socrates said of his own, became her guide in everything, her mentor, consoler, and primary physician. Estelle had a long conversation with Angeline about her own state. From this moment on her cure was assured, and we saw all the phenomena of ecstasy, catalepsy, and somnambulism develop rapidly in our little patient.

I will not describe her rapid development in detail. The account would be much too long and tedious. I will group these phenomena below according to their nature and principal character after the section on similar patients who also manifested these phenomena. But, I feel compelled to restate that within two days of Estelle's first magnetization on December 22, the first barely noticeable spontaneous movements in her lower limbs manifested themselves. The movement occurred only under the influence of magnetic passes. Estelle was barely eleven years old[1] at the time and still crippled after more than two years. The movements increased rapidly during the following days under the same influence. Six days later (December 28), Estelle sat up and shook her legs in the air one after the other, after rubbing them with newly fallen snow. She then raised them together to 45 to 50 centimeters above the sofa where we had placed her to be magnetized. On December 29, the electrogalvanic influence of gold manifested itself on her for the first time. The following day, while she was in a magnetic sleep, Estelle had scissors and a comb brought to her. With her eyes shut, she cut her hair, twisted it, washed up with as much—even more—dexterity than a clear-sighted maidservant with the help of a mirror and all the necessary toiletries. She continued her treatment with snow, despite her delicate chest, despite the spots on the entire periphery of the thorax that habitually tire her, and despite the cough that, for a long time, had made us fear tuberculosis of the lung. She put snow on top of her head and ate it, using a gold pin to take it very easily from the basin she asked us to bring her.

On December 31, Estelle was again put into a somnambulistic state with magnetism. She then washed up again and applied snow like before. Estelle then STOOD UP BY HERSELF AND WALKED. She announced, meanwhile, that these marvels of magnetism would stop when she would leave somnambulism, but that we mustn't worry since this extraordinary state would reappear with new magnetizations. This cure, however temporary it seemed, would gradually bring her to a normal state by pulling her out of the unnatural state into which her cruel sickness had plunged her.

She even announced her improvement would not be continuous and smooth. Instead, her progress would happen in fits and starts. Her cure would simmer in silence, so to speak, without being noticed much, until perhaps suddenly manifesting itself unexpectedly in three or four brilliant moments. She would herself be aware of this progress, thanks to something extraordinary going on within her. She did not yet know exactly how this would happen, but she would inform us of it at the right time. She added that we mustn't get impatient because "things couldn't go as quickly as we wanted." These revelations took place only in the crisis (somnambulistic) state but were forgotten when she awoke.

Finally, we saw the patient feel the electrogalvanic force of metals in a very obvious way. This influence increased rather quickly to the height of its power. This is a constant phenomenon in those magnetized or in analogous states arrived at spontaneously. I seize the opportunity to speak of these phenomena here, believing that they have not yet been described scientifically in a satisfying, intelligible way for those who have never witnessed them. I will expose, in some detail, what I observed.

From the beginning of my research in 1820, 1821, and 1822 on the very special nervous phenomena cataleptic patients manifest, I had noticed the unusual desire these patients have for gold, especially the purest kind. I had also noticed the clearly different effect on them of zinc, yellow copper, and magnetized iron. Before questioning these patients on the sensations these metals gave them, I wanted to run many tests to identify the relative consistency with which these remarkable phenomena emerged and evolved. I sought to know whether the phenomena derived from a law of nature not yet observed or whether they were simply the by-product of chance, whim, or patient ingenuity. As a result of this repeated investigation (about which I spoke to no one), I found that identical symptoms emerged when the same patients were placed in identical circumstances. I had to conclude that these phenomena belong to a natural order subject to positive, unstudied laws as immutable as the great laws that govern the universe. Therefore, it was necessary to study them. I undertook to do so.

At that time in 1822, I had two cataleptic patients, Annette Roux and Micheline Viollet. They were commoners who lived almost 40 kilometers from each other and were strangers to one another. I could see them at my convenience in Aix and Annecy, the two towns where I saw patients. For scientific purposes, I took advantage of this situation to exploit all the nervous phenomena each patient presented.

Annette Roux had bright red hair, freckled coloring, and slightly decaying teeth. She had always lived in the country, was lively, and perhaps even a bit overly emotional when annoyed. She was, however, accommodating in social relationships, frank, willing to do what others wanted, and kind and grateful toward those who helped her or were interested in her. She was subject neither to jealousy nor bitterness. She liked a clean, distinguished manner of dress and belonged to a wealthy family from the country. Micheline Viollet, on the other hand, was a large person with a sanguine[g]-lymphatic[g] constitution. She had long, silky chestnut-blond hair, and beautiful, remarkably white teeth. She was spirited in her retorts but had no social graces. She was capricious and emotional if her wishes were not granted. She was grateful toward those helping her only when absolutely necessary. Micheline despised money and luxurious apparel, putting little importance on her clothing. She spoke frankly and freely to anyone and about anyone. She was from a poor family in Annecy and earned a living working with her hands (as a seamstress).

Both young women, between 18 and 20 years of age, had become ill after the same kind of fright. They were religious, had good manners, and were not self-centered. They were menstruating normally without any noticeable disturbance.

I would do experiments on Micheline and then repeat them on Annette, and vice-versa. In the second and third year of treatment, I introduced them to one another and encouraged their friendship. From then on, I studied them in relation to one another. I must admit that I did not try to cure them during the first two years. I tried only to ease their situation as best I could. The phenomena I saw, however, appeared to be so extraordinary that I felt it was my responsibility as a doctor to study them well and thereby to serve others better afterward. I believed that I would rarely encounter circumstances so well suited to this kind of study and that I should take advantage of this fortuitous opportunity. With my knowledge and experience and these current circumstances, I began my experiments.

From the beginning of my research on nervous phenomena, as presented simultaneously by my two cataleptics, I noticed that when Annette Roux wanted to lessen intolerable pain on the top of her head, she rubbed the spot with a little gold pocket-watch she had requested. She always wore it suspended at her neck. Otherwise she would slide a big 100 franc gold piece into her hair.[2]

When Micheline Viollet was in crisis, she would eagerly seize all rings, pins, and gold change within her reach. When she put them in her mouth, I would observe the sudden, seemingly magical cessation of lockjaw, a principal phenomenon of the crisis state. A piece of steel placed between the teeth immediately brought on lockjaw, but gold always lessened it the moment the gold was applied. If yellow copper touched a limb, it became rigid.

These phenomena presented themselves in a consistent manner for a good while, even though my patients did not yet know each other. They could not have been acting in concert. Therefore, I had to conclude that the phenomena belonged to something real, natural, highly regular, but still unknown. The study of their laws was vital; it would be of great practical interest for the cure of nervous symptoms. Therefore, I questioned my patients on what they were experiencing and on why they were acting in this way. "Good question," I was told. "Don't you see, Doctor, that this helps me? If I am cold or if the top of my head is 'burning,' gold settles my nerves and comforts me. If my teeth are clenched, gold relaxes them. Rubbing myself with gold cures me."

At this time I read a summary of observations done in Turin by the academicians Avogadro and Michelotti (these observations appeared in the *Annales de physique et de chimie*). It concerned the electrogalvanism of metals or the galvanic power of each type of metal to produce electric currents. I had noted the rank-order these scholars had identified.[3]

I realized that galvanism or metallic electricity had an extraordinary effect on my two patients and noticed that the intensity of the effect was infinitely stronger than that of electricity produced by a machine. For example, I made a chain of 80 disks (copper and zinc) on a table, one-quarter of each disk overlapping its neighboring disk. If a patient in the crisis state put her index finger on the first or last disk of the series and if I had her touch the disk ending the chain with the index finger of her other hand, she felt—at the moment of contact—an electric shock pass from one hand to the other as well as across her body. The shock was much stronger than the shock from an ordinary magic tile,[4] 30 centimeters in diameter and strongly electrified.

Another experiment confirmed this one. I placed the outer armature of a Leyden jar at one end of the chain described above. I hit the other end of the chain with the mouth of the bottle to see if the bottle could be charged enough to obtain a noticeable shock. What I suspected happened exactly as expected. I obtained such a jolt, however, that my patients said that they had never felt one like it. (Once the bottle had lost its charge, it no longer caused any shock.) No one out of crisis felt anything similar during the test. We must rely on the reports of patients about the feelings they experience. I have repeated the experiment so many times, however, that I have no doubt about the veracity of their statements.

I also observed that patients entering into crisis at the dining table took great care never to touch their knives where the handle met the blade (if composed of two different metals.) While in crisis or somnambulism, my cataleptics never grabbed a knife with a silver, copper, or tin handle except by either the handle or the blade. If, by chance, a patient happened to touch a knife where the handle and blade met, the patient dropped it immediately. She would shake her fingers, looking at them in surprise, as if wondering what had happened. When I asked about the feeling that was strong enough to surprise them that way, they responded that it felt as if a spark of fire had come out of the knife and burned their fingers. These patients were very careful not to touch a watch key of gold or silver where the square of steel met the body of the key. When they needed to use the watch key, they wrapped it carefully with linen or another fabric.

I was struck that these events, if under identical circumstances, always happened in this way. I could not doubt they belonged to a normal order of natural phenomena previously undetected. I resolved to push my research further. I got some large disks of the metals named in the *Annales de physique et de chimie* and gave them to Annette Roux. I said, "Annette, here are some different metals; let's see what effect they produce on you by flicking them with your finger."

I had her put one of these disks on its edge and press the middle finger of her left hand on the highest point. Then, pushing the edge of the disk toward the right hand with the middle finger of the left hand, the patient managed a rotation of the coin on its vertical axis. At the same moment, she let it go with a gasp of surprise. When I asked about what had happened she said, "I felt——I felt a spark on the finger of my right hand where I had flicked the coin and then I felt an inner tremor go through my left arm from the finger that held the coin all the way to my heart." Perhaps there is no need to repeat that I did these experiments numerous times, but only when the patients were in a crisis state and always with the same results.

Since, however, I had disks of zinc, silver, iron, tin, lead, yellow copper, and red copper, I successively subjected all the metals to the same experiments, always noting the results. Annette Roux taught me that each metal produced noticeably different results. The electrical sensation extended to the middle joint of the finger in one and to the wrist in another. In one experiment, the spark had been so strong that it was felt to the elbow; in another, it had been much weaker.

Because it was difficult to hold the round disks on their edge vertically in mid-air while turning them, I decided to create a kind of axis by implanting two small points of steel at the extreme ends of a diameter of the disc, and I began my experiments again. The results were no longer the same. The percussive effect on the disk, suspended like a sphere parallel to the horizon, set in motion a much faster movement. Consequently, from the first movement of the coin, there was a significant electric shock traveling from the finger turning it toward the finger suspending it. Annette immediately stopped this tiring exercise. She carefully examined the coins used in the experiment and looked at her fingers with surprise. She shook them and then began the same operation again several times, but with more care. She was attempting to see the effects of the rotation of this unusual new electrical machine on her. Soon, however, what had surprised and startled her became an amusing game.

Meanwhile, since a gold disk the size and thickness of other disks would have cost me at least 250 francs, I did not have it made of pure gold. I believed I would get the same results with a disk of gold-coated copper (four layers of gilding). Thus I had two made at one of the best goldsmith-jewelers in Geneva who gilded two sequins from Holland. During her normal state Annette had found it very pretty and expected marvelous results from it. I was greatly astonished, however, when she said in somnambulism, "Oh! Monsieur Despine, you have been tricked if someone sold it to you as gold. Yes, you have been scandalously tricked. You have been robbed, for here there is neither the weight nor the value of pure gold. If there is any gold, there is very little, for the piece 'burns' me almost as much as the others when I touch it, yet it doesn't comfort me like my gold piece."

This took place in December 1823. From then on, all cataleptics who came into my care manifested the same phenomena, with some nuances. Several patients who admired silver-coated pieces rejected them disdainfully when they became somnambulistic. The same thing happened for diverse small accessories such as rings, necklaces, bracelets, earrings in crysocale, in false gold, or in gold-plated copper. In the patients' waking, normal state the beauty, purity, and brilliance of the coins had impressed and fascinated them. But if the patients were in the crisis state, only the intrinsic, metallic value of each piece and its related electrogalvanic power determined its value. I am also convinced that the religious associations some patients attached to rings or devotional medals that had been blessed made no difference in how much the objects were appreciated. When in crisis, the patients always preferred the rings that were heaviest in weight, mass, and size—whether or not they were blessed. These soothed them best.

In these experiments, I have always been struck by the consistency in the classification of the metal disks. The patients chose a sequence that generally corresponded to the sequence identified by Avogadro and Michelotti (I could even say always, rather than generally, since there were so few exceptions). Yet gold occupied the extreme negative of the chain and zinc, the extreme positive. In between gold and zinc came silver, copper, iron, tin and lead.[5] The patients placed platinum immediately next to gold but only after the gold and never before it. Yellow copper and alloys have always tired patients greatly. Their place in this galvanic scale varies considerably, since it depends on the nature of the alloy and on the specific combination of metals in the alloy.

A timepiece such as a watch, for example, provoked very special effects in Annette Roux and Micheline Viollet (it did so in others later on, as we saw in Estelle's account). I classify these effects among the galvanic phenomena described earlier. My two patients had more energy when a gold watch was suspended at their neck or elsewhere by a gold chain. (The gold chain could be replaced by ribbon or string. Silk should never be used. Contact with it, as with glass, fur, and other idio-electric bodies, gave the patients a very strong burning sensation). If the watch were wound and ran well, my patients did not fall into a faint. As soon as the watch stopped, however, they immediately fell into syncope.[g] If there were glass on the watch, the patients, even with their eyes closed, reached carefully so as not to touch the glass. A gold watch was preferred to a silver one and to a watch whose case was of copper, crysocale, Manheim gold,[6] or any other composition. A smooth watch casing was preferable to a case with embossing, engraving, enamel, and gold made of different colors. Such were the phenomena I observed in Annette Roux and Micheline Viollet in 1822 and 1823, and in Estelle, 1836 and 1837. Henriette Bourgeat and Madame Schmitz-Baud from Geneva manifested the phenomena as well.

The constancy of these phenomena in my experiments proves that they always manifest themselves in the same way in patients of this kind. The cause of the phenomena is clearly connected to the galvanic action of metals. A watch is a system of movements composed of copper, iron, or steel pieces. Some parts are made of an ordinary metal whereas others are modified by gilding. As soon as this system of diverse metals is set in motion, a galvanic power, much more marked than when the watch is not running, immediately results. This happens after the wheels of copper, the steel pinions, the axis of the wheels, and the copper-plated platinum on which they move rub against one another. Regularizing the overall movement is the traction of the spring on one side, resistance of the chain on a cylinder in a different metal on the other side, and the general equilibrium of the mechanisms. When all of these mechanisms move, patients react in obvious ways. In crisis, the patients' impressionability to a watch is much greater than if they are in the normal state. The patients' sensitivity to the mechanisms in a watch is especially believable, since touching the juncture of two metals on a pocketknife or watch-key caused them to feel a small electrical shock. Moreover, facts cannot be denied, especially when they concur with our present knowledge in physics by worldwide observations throughout time. It is indisputable that nuances of phenomena analogous to those I describe have presented themselves before. Historians had been unable to explain them. Now, though, they are more understandable to those who reflect on them (see my endnote 15).

Patients are extremely sensitive not only to the electrogalvanic power of a watch. Like Estelle, they also respond to the rhythm of the watch's movement. A watch whose movement was perfect was preferred to one whose movement was good but not perfect. Madame Schmitz-Baud, a good musician whose work involved putting finishing touches on musical pieces at rehearsal, rejected disdainfully any watch having the smallest irregularity in its movement. In 1816 in Wilna, the famous professor Joseph Franck applied his watch to parts of Louise Baerkmann's body.[7] She was immediately less lethargic; she perceived the motion and rhythm of his running watch.

I often saw the movement of a watch and music box support the vital movements of Annette Roux, Micheline Viollet, and Henriette Bourgeaut. The vibrations of these objects imitated the sensations of vital human functions that otherwise would have stopped. Had they ceased, these patients could have died. Estelle knew nothing about events of this nature in my practice. Nevertheless, such things happened to her in the same way. Was the watch working? Estelle "was working" too. Her somnambulism became active, and she enjoyed all her vital movements with ease. Had the watch stopped? Life seemed to suddenly stop too. Some say these effects are imagined. Some say most phenomena manifested in patients with nervous illnesses are imagined when the effects are irregular and cannot be explained anatomically or physiologically. Nothing, however, can be explained by the imagination. The difficulty of understanding a condition is only put off, not resolved. I ask these critics, "What do you mean by the imagination?" I doubt that their response will satisfy me! However much one wants to attribute to the soul the more or less fantastical images observed with nervous phenomena, galvanism explains most electrometallic phenomena I have observed.

For the same reason, patients prefer the harmonious music of wind instruments and those with catgut rather than music of metallic-stringed instruments. When in crisis, they also carefully avoid big silk guitar strings covered with silver. For the same reason, the harmonica and organ affect vaporous women.

These phenomena are very curious and worthy of research today by physicians or physiologists. I will return to this subject later by treating the very remarkable acoustic phenomena several of my patients exhibited in crisis.

Endnote 4

Hysteria and Its Diverse Phenomena

As I explained in Phase One of the Magnetic Cure, after six months in Aix, Estelle in the waking state could barely put a foot on the ground or even take a step without hurting herself or fainting. Symptoms often seen in emotional and nervous patients prone to hysteria are both a tendency to faint when trying to sit or stand up and a weakness in the legs (a kind of partial paralysis) expressed in increased sensitivity and decreased mobility. These patients are usually women. I have frequently seen these phenomena in young people approaching puberty and in nervous, high-strung women who are somewhat older and who live in considerable comfort and luxury. I have never seen these phenomena, however, in post-menopausal women.

The phenomena of hysteria include bad headaches, ordinarily centered on the top of the head or located on a parietal crest. Patients frequently think of a bad headache as a migraine and call it "rheumatism of the head." They also experience extreme sensitivity of the skin, localized on the spine or the sternum, and sometimes on the abdomen, hips, or hypocondrium.[g] Complaints may include stomach problems, irregular menstrual cycles, and persistent constipation. Ordinarily, the gait is unsteady as well. Patients say they are "lopsided." It seems to them that their pelvic bones are "enlarged and vacillating at the joints." Finally, these patients report a peculiar, very tiring sensation of internal tension and warmth in the pelvic organs. This state is painful and disquieting, even though it is more a condition of swelling than of inflammation. The patients, however, believe that the core and source of all their pain and varied ailments is in the uterus. They imagine themselves ill with acute or chronic metritis[g] that will soon degenerate into tumors, ulcers, or related conditions.

When this illness affects the breasts, as it often does, the patients do not talk about the condition. They remain silent—embarrassed or fearful that the diagnosis will be much more serious than an illness of the hypogastrium.[g] The breasts may feel swollen and tender. If, on the other hand, the pain is intense, it seems throbbing and jabbing. At times, the discomfort is in the whole breast; other times, only the nipple and surrounding tissue are affected. This condition, nevertheless, is only "sympathetic"[1] (or, at least, I have always found it to be so), despite the fact that the slightest pressure is intolerable and that the breast is sometimes darkened or marked with black streaks. In several cases, I saw the coloration disappear instantaneously when treated with electrical sparks or small charges. If the tissue had been even slightly diseased, this treatment would have failed.

I put Estelle's illness within this category of disorders, even though puberty had not yet begun. She had experienced some pain on both sides of her upper chest and occasionally experienced slight swelling. A few of the phenomena mentioned above were also evident. I had no doubt that her affliction was essentially hysterical, even though her symptoms first appeared to be organic and not functional. I can say the same for Sophie La Roche, Henriette Bourgeat, Annette Roux, Micheline Viollet, and for countless other young people between the ages of 15 and 20. The details of their illnesses and presenting phenomena vary only slightly.

Bloodletting, very often used to treat these illnesses, brings relief, but the physician can unwittingly overuse this intervention to a patient's detriment. The forced and unnatural depletion of blood disturbs the functioning of the whole circulatory system. Indeed, while bloodletting initially improves the state of the patient by decreasing the amount of blood, in the long run it increases circulation and causes tumescence in the unhealthy organs by drawing blood to them. Consequently, there are rigidity, swelling, and pain, but neither inflammation nor a true phlogosis.[g] Warm, prolonged sitz baths, a light and refreshing diet, and similar treatments effect a more certain, lasting cure. General or local bloodlettings by lancet or leeches should be used only as a last resort and only when the patient is suffering so much that temporary relief is preferable to continued pain.

From the beginning of her stay in Aix, I suspected that Estelle, despite her young age, suffered from an illness of this nature. I treated her slowly and methodically in order to observe what would happen and soon became convinced that my assessment of her condition was correct. As her symptoms unfolded, I no longer considered her almost complete paralysis to be an ordinary idiopathic paralysis resulting from a compression or internal bleeding acting directly on the brain or on some part of the central nervous system. I saw her condition to be a "sympathetic neurosis"[2] and believed her problem to be psychosomatic in origin. Estelle's frequent headaches, cough, tightness in the chest, palpitations, delirium, and other phenomena would have been diagnosed as inflammatory, or at least sub-inflammatory by many doctors who followed Broussais.[3] Thus, I used no bloodletting, leeches, or purgatives[g] to treat this patient. The baths alone, along with a healthy routine and diet, were enough to cure Estelle.

In my current formulation of this kind of psychosomatic paralysis, there are both anatomical and physiological explanations. Weakness of the legs appears to be caused (1) by increased blood flow to the organs at the base of the nerve trunk and (2) by a lack of appropriate innervation[g] caused more by an uneven distribution of nervous fluid than by insufficient fluid. It is difficult to believe that either condition alone could cause paralysis. But it is easier to believe that a very impressionable individual with both conditions might exhibit all the phenomena cited above.

In the first point of my hypothesis, when patients with this weakness sit or stand up, there is a natural increase in blood flow. For the second point, I do not attempt to explain the uneven distribution of nervous fluid by purely anatomical relationships. I refer instead to certain physiological phenomena always seen in these kinds of paralyses. (These phenomena are never seen in paralyses resulting from compression either on the encephalon[g] or on the conductors of nervous fluid that radiate from the brain to all points of the body.) Finally, through inductive reasoning, I will try to draw satisfying, plausible explanations for the cause and effect of these phenomena.

In psychosomatic paralyses I observe that "vegetative" life always continues, despite the often complete disappearance of sensitivity and mobility. Flaccidity of the flesh and emaciation of the muscles—conditions that show up in ordinary paralyses—are never present here. In functional paralyses, the limbs continue to be nourished and develop more or less as before, since nourishment and blood circulation are neither suspended nor stopped in any noticeable way. Normal blood flow resumes with good health. Until this happens, however, the patient has a very noticeable sickly, skin tone. One could compare this state with that of sheltered plants in dark greenhouses or cellars during the winter. The plants become discolored because of absence of light, although they continue to live and vegetate. Color quickly returns, however, when they are exposed to bright sunlight and to the tonic, salutary effects of fresh air. Similarly, everything in the body returns to normal as soon as the nervous fluid has regained its normal flow and ordinary distribution. This return to health happens INSTANTANEOUSLY. Likewise, paralysis and related phenomena reoccur immediately if something disturbs the flow and distribution of this same fluid.

I believe that, by virtue of an immediate, singular action of electricity on the nervous fluid, electricity has its therapeutic use in the treatment of neuralgia and other illnesses of the nervous system. Electricity does not act, however, by increasing or decreasing the mass of fluid in the body. Indeed, electrotherapy redistributes the fluid so that it is balanced throughout the body in a uniform and natural manner. Following my formulation, therefore, electricity, on its own, often RELIEVES AND HEALS neuralgia by correcting the maldistribution of the nervous fluid or by hindering its abnormal accumulation at one part of the body to the detriment of other parts. Thus, neither strong sparks nor shocks suit in these cases. (They seem to work in ordinary paralyses, however, following apoplexy[g] or compression.) Instead, little sparks (aigrettes[g]), and the electric brush and bath[4] are suitable. In these psychosomatic conditions, physicians must primarily attempt to distribute this fluid in a uniform manner since the nervous fluid in unhealthy cases does not follow the natural, ordinary laws of distribution. For this reason, the psychosomatic condition demands a much milder application of electricity. The same can be said for animal magnetism, which will be particularly successful in these cases if applied with care and gentleness. Like electrotherapy, magnetic treatments can succeed in these cases only by acting on the vital or nervous fluid.

Endnote 5

Imitation, Echo, Attraction-Repulsion

The phenomena of imitation observed in nervous maladies are very curious. We find traces of this ability to imitate in countless ordinary daily actions. Several disciples of the most famous ancient philosophers unconsciously assumed the tics, allure, and expressions of their masters simply by living with them. Socrates' disciples were criticized for imitating even his faults.

I do not discuss this particular power or ability to imitate here. Instead, I study the impressionability in those with nervous disorders that compels them to repeat movements others do. I have identified two types of impressionability. The first type includes phenomena of imitation while patients are aware. Moreover, a patient and others can, to a certain point, influence this behavior by will or emotion (terror, joy, fear, unexpected pleasure). Within this group falls the contagion one usually feels when another person yawns, even if the yawn is faked. The effect on others is to yawn. Boerhhaave reported a similar case concerning a boarding school of young ladies in Harlem.[1] A student had an attack of nerves resembling epilepsy in front of her friends. She caused them a great fright, making a very unpleasant impression. The next day, when this young woman had another attack in front of the same persons, all experienced an identical attack of nerves at the same time. This phenomenon also manifested in our little Estelle when she witnessed in a child a brief episode of St. Guy's Dance.[2]

The second type of impressionability—distinct from the first type—identifies the phenomena of imitation while patients have no awareness or feeling, so their acts are purely automatic. Neither the patients' will nor imagination can influence them. I observed it for the first time in 1822 while treating Mademoiselle Annette De Roussillon. I called it speculative imitation since it manifested itself with all the spontaneity of the best mirror. I recorded the case in my report to the government on the medical season of the springs in Aix for that same year (1822):

> In the state of lethargic sleep, Mademoiselle Annette De Roussillon presented a very unusual phenomenon I have never seen reported anywhere. I call it speculative imitation, since it is comparable to the effect of a mirror.
>
> She was lying down in bed. I put my right hand on the patient's head to establish a relationship with her by the occiput.[g] If I moved my left hand in the direction of her right hand, her hand came closer to mine, following all my movements as if in a mirror. She performed the different movements precisely as I did them, however bizarre they were. This imitation stopped as soon as I took my hand away from the occiput,[g] or if the modeling hand moved away more than 60 centimeters from the copying hand (even if my other hand remained at the occiput[g]). The copying hand remained where the modeling hand had left it when its influence had ceased, as if waiting for the modeling hand to come back to draw it away. It fell only when the first hand (my right hand, in the example cited) left the occiput.[g]

In more or less the same way, I then observed this curious phenomenon repeated in all my cataleptics. Estelle, however, who was observed closely by her mother and me (we never lost sight of her), presented this behavior much more frequently than the others did. When the patients were asked about the cause of these phenomena and about what was happening to them during these episodes, all of them, with the same tone and manner, said, "Sir, I do not know!——but I do not want to do anything——a force that I must obey pulls at me——I obey this force, despite myself——It seems that a part of my body no longer belongs to me——I know that I am doing something——I feel it——but I cannot say what it is——I have absolutely no idea." This is quite curious and remarkable.

In most of my patients, I have also observed the phenomena of echo as well as attraction and repulsion. These phenomena seem to belong to the same physiological cause as imitation because (1) a power against which most patients can offer no resistance determines the phenomenon; (2) there is usually no awareness of what has occurred; (3) the phenomena stop the instant the material or mechanical action that caused them stops.

I call the echo phenomenon a patient's mechanical, unconscious, and verbatim repetition of all that is said by those around her. She does not know what she is saying and has no memory of it. She repeats good things as well as bad—silly, stupid statements as well as the most elevated thoughts. She does not laugh, blush, or cry. It is like an echo off a rock that sends back what it hears by acoustical laws without having any awareness of what it is doing. Feelings have nothing to do with it.

Endnote 6

Various States of Somnambulism

It is rather difficult to establish clear-cut divisions between the diverse phenomena catalepsy presents. The reason for this diversity is simple. Each individual manifests noticeable differences in the overall symptoms of the illness, as well as in the nuances or modifications of each symptom.

In my 1822 report to the government I wrote about six very distinct states in Mademoiselle Annette De Roussillon:

- spasms and convulsions;
- catalepsy;
- lethargy, syncope,[g] or apparent death;[g]
- apoplectic sleep;[g]
- partial state of crisis that gave the patient an inner sensation of a double life;
- somnambulism.

The following year I was able to consolidate the six diverse conditions of the patient into four states and identify the four overarching principles:

- The waking state. This was the patient's ordinary, natural state upon arrival in Aix. During the waking state, her emotional and physical way of being was genuinely sickly and totally different from her way of being during the crisis state. In fact, when in the waking state in Aix, Mademoiselle De Roussillon was completely different emotionally and physically from the waking state before the illness started.

- The state of spasm and catalepsy.[1] This state of nerves impacted especially on the muscular system and naturally subdivides into as many types as it has different forms.
- The state of active somnambulism. I subdivide this state into as many varieties or types as the diverse scenes it can represent. These include scenes of terror, as well as visions, hallucinations, rapture, ecstasy, and mimicking.
- The state of dead or passive somnambulism. It consists of syncope[g] and its four expressions (fainting, lipothymy,[g] lethargy, and apparent death[g]). The phenomena of imitation, echo, attraction, and repulsion belong to this state.

Estelle presented all these diverse states. Later I will provide more details of the curious phenomena presented by Sophie La Roche, Henriette Bourgeat, Alexandrine Guttin, and other cataleptics or victims of magnetic crises whom I treated after Estelle.

Endnote 7

Patients' Preferred Vegetarian Diet

A Summary of this Endnote: *During Despine's time, medical experts assumed that meats were more easily digested and provided more nourishment than vegetables. Patients with neuropathic afflictions (such as hysterics, cataleptics, and hypochondriacs) preferred a vegetarian diet because, in Despine's understanding, the nervous disposition of the digestive tract functioned like an excess of electricity. Fatty, greasy foods—including meats, fried foods, pastries, alimentary substances, very sweet tonic drinks, and even chocolate with its base of cocoa butter—"burned" their stomachs, constipated them, and aggravated their difficulties. Despine was convinced that Nature spoke loudly through the patients' instinctive appetite for a vegetarian diet and that their doctors must honor this preference. Sometimes somnambulistic patients prescribed their diet. Estelle was more resolute about her diet than any other of his patients, undoubtedly because, as Despine reported, she felt tricked in Neuchâtel—her asparagus had been cooked in beef broth. During her recovery, Estelle eventually ate some ham with pleasure.*

Endnote 8

My Professional Reputation

Rumors have been spread publicly, either maliciously or foolishly, about my madness and stupidity as well as about my thirst for fame. Two amusing anecdotes will set the record straight about what usually goes on in thermal establishments. Aix, however, is a class apart. In fact, physicians practicing medicine here during the summer are on good terms. They understand and help one another and communicate amicably their observations of the rare, curious cases in their clientele—all in the interest of science and the sick.

This pragmatic, collaborative approach was introduced by my excellent father, Joseph Despine, the first doctor the government appointed for the baths. The approach has not changed since the earliest days of his tenure. We hope that, for the sake of the patients and for medical progress, it will never cease. The situation is different, however, for places where visitors lodge. In such places, each establishment has its own doctor whose sole reason for practicing medicine gratuitously during the whole year is to have a clientele during the season of the waters. Although that practice is understandable, I applaud the hotels where visitors are encouraged to use doctors recommended by their own physicians or acquaintances (rather than use a house doctor). This is not, however, what happens in places where industrial resources are only seasonal. In these cases, pride, self-interest, and cliquishness prevail. Self-satisfaction, very natural to those who have never traveled, makes it impossible to imagine that anything in the world is more beautiful than the "rooster of their own belfry." Indeed, everything at home is perfect. Such provincialism may also arise from jealousy of what strangers in our town do, even if there are obvious benefits to their work. We prefer to get credit for things ourselves. People who believe themselves scientific want to be the first to say everything, do everything, and discover everything. But discoveries are made only little by little, especially in the biological and positive sciences.

Now for the two stories.

The first anecdote concerns the Marquis de G., who told me the story himself. It happened a few years ago. I will not say which year for fear that some mean-spirited, overly curious foreigner might delve into our printed records to discover who figures in this drama and do the Marquis harm. It is enough for me that this person recognize himself in my story and that any foreigners who read it know how they may be sometimes intentionally misled. Meanwhile, I reserve the right to give the Marquis de G.'s address to anyone who doubts the truth of my story and needs an eyewitness account.

The Marquis de G. arrived from Paris with his young wife. He came for himself, having been recommended to me by Madame la Comtesse C., whom I knew. Upon his arrival in Aix, the apartment he had previously reserved was occupied. Therefore, he was lodged in a bigger, comfortable home that, however, looked out onto a little courtyard. Shared by several other apartments, the courtyard was dreary, loud, and dirty. The couple arrived at night and settled in as well as possible. The next morning, though, the Marquise fell into the blackest melancholy and wanted to change lodgings (she was used to beautiful lodgings in Paris with views of the Champs-Elysées). Her husband, accustomed to military life, was bothered little by it, but suggested to the proprietor a few small changes to make the place more comfortable. Although the changes were possible, the proprietor refused to accommodate the Marquis and his wife. Consequently, the Marquis immediately sought better lodgings. He found a place and moved there quickly to please his wife and finish his treatments. This worked out marvelously.

Feeling particularly happy on the day of his departure, the Marquis expressed a desire to purchase a place in Savoie. (He was very content with his treatment and knew his wife was unhappy about leaving the charming view from their new lodgings.) He asked me to inform him if *La Maison Chevally* were put up for sale and added, "Monsieur Despine, there are some very mean people in Aix."

I responded, "In Aix, as in all other little provincial towns, there are good and bad people. In Aix there are many more good than bad. But, Monsieur le Marquis, why do you bring this up? Are you unhappy with the town?"

Photo 12 Town dwelling of Antoine and Péronne Despine in Aix-en-Savoie at 12 Place Carnot.

Source: Courtesy *Archives municipales d'Aix-les-Bains.*

"No," he responded, "I leave very happy with your waters but, my dear Monsieur Despine, I must tell you something in confidence. People disapprove of you here, although wrongfully so. What I have heard and seen myself proves to me that this establishment owes all its improvements to you. I declare to you that the impetus you give it will take it far. Many people, however, do not give you the credit you deserve!"

I responded to what was flattering in the Marquis's explanation as politely as I could (the conversation had taken place in the presence of his wife), but I was uncertain what to say. "I care little about what people say about me or think of me, Sir, as long as I have fulfilled my responsibilities. But, tell me, what upsets you so much about Aix and its inhabitants?"

"When I arrived from Paris," he responded, "you know that I went to your home quite late. I found you at your desk, working by the light of your small lamp. You were not out socializing at the noisy casino. Because of a short conversation I had with my proprietor, I went directly to your home instead of asking you to come to mine. I had asked her to contact you about coming to see me the next morning."

" 'Monsieur Despine is not the house doctor,' said this woman. 'If you desire a doctor, I will contact our doctor, with whom you will be very happy. You apparently don't know that Monsieur Despine has gone crazy. He is old, no longer gets around, and spends most of his time in his country home in Saint Innocent.' "

"I asked her, 'Crazy, Madame, since when?' "

"She came back with, 'Oh, Monsieur le Marquis, for a very long time.' "

"So I said, 'I don't understand. He wasn't crazy two weeks ago. The Countess C. in Paris advised me, 'You are going to Aix? Go to Monsieur Despine. He's an old friend of my husband's and my family's. Although he was not my doctor when I went to Aix with my father, you will be happy with him. He is aging, but he is not a schemer at all and has dedicated his whole life to the waters. We often have news of him through our doctor in Paris with whom he corresponds.' "

" 'You see,' I said, 'Madame, he was not crazy then.' "

"The woman stammered a few words of vague apology and added, 'Monsieur le Marquis, I don't know what to say. You believe yourself better informed than I am, but I can assure you that Monsieur Despine is believed to be crazy in Aix. For many years he has treated only madmen and madwomen and has dealt with animal magnetism, somnambulism, imaginary illnesses, and other things like these.' "

The Marquis de G. continued, "I responded to my hostess, 'Well! Madame, I will go see for myself what's going on. Please get someone to drive me there. I will judge for myself.' That is how you are treated by your compatriots, Monsieur Despine."

"But," I said, "I am not from Aix, but Annecy; circumstances have brought me to Aix."

The Marquis responded, "Oh, I understand that you are not from Aix. In any case, watch out for these inhabitants of Aix, especially those who flatter you the most." I thanked the Marquis de G. and continued to do good to my detractors and others when opportunities presented themselves.

The other anecdote has to do with my excessive ambition. People imagine I have a thirst for fame—to make a name for myself. I was accused of such in *Essai analytique-médical et topographique sur les eaux minérales, gaseuses, acidules et thermo-sulfureuses de Lapperrière près de Moutiers en Savoie* by Joseph Marie Socquet (1824, p. 108).[1] Since I am not named, this little malicious act went unnoticed by most readers and even by several friends who read the article but understood nothing in it. The author (whom I used to know well) doubted the mysteries of animal magnetism, somnambulism, and catalepsy. He asked to see my patient Micheline Viollet[2] in crisis (he was passing through Aix in 1823). I was pleased to show her in this state. I had done so for colleagues living in Aix and visitors passing through. I wanted him to avoid useless research and tests or experiments that might upset Micheline and hinder the frank manifestation of the most curious phenomena she manifested in her crises of somnambulism and catalepsy. So I spoke to him of the principal phenomena described very well in *Amélie, ou Voyage à Aix-les-Bains et aux environs* by the Count of Fortis (1829, vol. 1 of 2, p. 195). He wished to see her alone, without witnesses. Full of confidence in this scholarly, esteemed colleague, I went about my business. I do not know what happened during my absence, but upon returning I asked, "And so, my dear Doctor, what do you think?"

"Not much," he said to me, "This girl is tricking you. She tried to trick me too, but I quickly pointed out her mistake. At first, I concurred with you and Dr. Vidal that, at times, there was transposition[g] of the sense of hearing to the feet, neck, and epigastrium.[g] Don't trust it, I tell you, don't trust it. This girl wants to make a public sensation. Like all young girls, she wants to get herself noticed and then catch the first young man who shows interest in her. All that, believe me, is nothing but charlatanism and trickery."

"That may be, my dear Doctor," I responded, "but you wanted to observe without any witnesses, including me. Therefore, I can say nothing about what you saw or did not see. In any case, do you really believe Micheline Viollet could have faked these symptoms for everyone in Annecy, Geneva, Chambéry, and Aix who observed a crisis state in her? I know that mischievousness is characteristic of patients of this kind. Here, however, we see either new nervous phenomena or little known phenomena. They are related to magnetism as we know it today and to what we hear about the Brahmas of India, Tarentism,[g] Tigretier,[g] the Abyssinians, and the ancient Oracles. You know that long ago the *Académie Royale de Médecine de Paris* prohibited magnetism. Despite that, magnetism has come back stronger than ever! It seems to me, then, that it is in the best interest of science and medicine that doctors study these phenomena. If there is, indeed, something here and if magnetism is a useful therapeutic method, then clearly doctors should use it. Magnetism should not be left to the whims of the uninformed or to amateurish enthusiasts whose interest will quickly pass. On the other hand, if there is nothing to magnetism, it must be rejected completely. But, after all, it seems to me suitable and prudent to examine this question objectively without prejudice. We must multiply observations and explore them in depth to obtain clear, trustworthy results."

"Ridiculous! Ridiculous!" my colleague said to me, "the situation has already been judged by scientists wiser than we are!" In the same vein, Dr. Socquet wrote that fairly virulent note against me I mentioned above. I referred to his note in my *Rapports annuels au gouvernement sur la saison des eaux*.

I did not deserve this personal attack. The attack was made only because, in my annual report, I gave honorable mention to Charles Gimbernat's discovery in 1822 and 1823 about the formation of albumin in our mineral waters. I imprudently showed it to Socquet upon submitting my last report on the medical season of 1822. I swear before God that my intention in announcing Gimbernat's discovery was not to disparage the chemical knowledge of my colleague, Monsieur Socquet. In discussing the history of our baths, however, I had to mention the curious observations of this renowned but unfortunate visitor (Monsieur Gimbernat) who spent several months in Aix to study our waters. He suggested excellent ideas that we have used to improve our baths. His findings were extremely curious but completely peripheral to the analysis of the waters in Aix done by Professor Socquet in 1803. Gimbernat's observations on the spontaneous and extemporaneous formation (as Monsieur Socquet calls it) of these products have been confirmed since then by Messieurs Berthier, Chevallier, Anglada, Guibour, and Fontan. Our learned chemist Gimbernat advanced their work (see J. Bonjean, 1838, pp. 38, 114, 132, 213–214, 225), leaving Carrère, Lavoisier, Fourcroy, Vauquelin, Chaptal and other illustrious scholars far behind. Each of these scientists knew much more than his predecessors but did not know everything we know today.

Endnote 9

The Impact of Extreme Emotional Experiences

As I wrote in Phase Two of the Magnetic Cure, the emotional irritation and exaltation Estelle feels following a spirited confrontation momentarily aggravates her problems. She sometimes goes into a rage. Her intellectual functioning is troubled. She makes no sense; it is as if she is having a fit. This state was foreseen by her in somnambulism. Her sensitivity becomes excessive and she suddenly develops a response to colors. The problem is at its worst. Then Estelle tells the entire story of her illness spontaneously in somnambulism, with a coherence and intelligibility of which she was incapable in the waking state.

I have grouped Estelle's phenomena into this endnote so the reader might appreciate

- the negative impact of emotional upset and domestic quarrels on those suffering from nervous problems (especially those affected by emotional irritability). The degree of reactivity seems to be directly proportional to the intensity of the emotional upset and to determine the number of observable unusual phenomena. These phenomena not only overlap but also overarch one another so that the more extreme presentations are superordinate to the lesser presentations;
- the significant, rapid development of all the unusual phenomena which can bring patients to such a serious point that rage, despair, folly, and even death can result;[1]
- the heightened development of intelligence in the patient. (A prudent, wise doctor will always try to limit what patients say and do

> in this state. He will never mock them. For, just as a "sword wears out the sheath," patients made to "perform" become exasperated, thinking a spectacle is being made of them and that they are an object of curiosity rather than care. They are especially angered to see they are not taken at their word.)

Several of my patients in crisis have told the story of their illness and treatment with method and extraordinary discernment. They identified the causes of their pain and mistakes in the treatment and suggested what should have been done differently. For the moment, it is useful to know that when these patients are in crisis, and especially in the state of syncope,[g] they find themselves reduced simply to inner consciousness. They are completely isolated from the outside world by abolition of their senses. Consequently, they are deprived of the ordinary means of relating to others. In the crisis state, therefore, they seem to withdraw into themselves and to be concerned only with their own problems and treatments. Whenever I ask what they are thinking about, they all respond, "Myself!"

Responsiveness to colors is another phenomenon worthy of note. Poppy red put our young lady from Neuchâtel into crisis. Annette Roux was put into crisis one day in a public carriage because a traveler had an umbrella of red crimson silk (this occurred even though the umbrella was hidden from view). No one except the traveler knew the red umbrella was in the vehicle. Annette herself pointed it out when the driver asked her why she had experienced an unexpected crisis (he was touching her fifth finger, where she most frequently experienced transposition of the senses[g]). Violet has always greatly fatigued all of my patients. This event seems to belong in part to the class of galvanic-metallic phenomena.

Estelle's illness had been dormant for a long time, manifesting itself in the form of a slow, gradual paralysis. Magnetism changed this dormant state, intensifying it gradually. The illness soon culminated, then declined little by little and finally healed completely. Asked when her long illness would end, Estelle always responded, "It will be long, very long. I'm moving toward healing, but slowly. Don't be impatient." My other patients whose afflictions resembled Estelle's responded similarly. None were mistaken! Could one not conclude that when the malady is advanced and well established, it must pass through certain phases before it can be cured? (This is true of most acute illnesses.) The doctor's great ability is to remedy the situation by promoting symptoms, but never beyond a certain point. He must also facilitate passage through the phases, repressing all that is eccentric or strange and using the patients' characteristic power of instinctive intuition that allows them to identify the most advantageous treatments. He must also avoid deliberately provoking the extraordinary phenomena debated purely out of curiosity in public. (There are so many conflicting reports about somnambulists and magnetic crises.) The physician must be content to observe the phenomena when they manifest themselves and make the phenomena work to the patients' advantage for healing.

Endnote 10

Patients' Distinctive Language

As I wrote in Phase Two of the Magnetic Cure in Aix, Estelle's ball bursts and, from this moment on the illness declines, just as she had predicted. Each patient with a disorder like Estelle's uses unique expressions to speak about herself, her pain, and her cure. We often laugh at their diction because the words are funny and unusual. Estelle speaks here of a "ball"—not to be confused with the *globus hystericus*[g] that so often tires hypochondriacs and patients with nervous disorders. Estelle was not referring to this problem, since she had no abdominal swelling, choking, suffocation, visible lump, or gurgling. She could find no better word to describe this sensation than "my ball."

Annette Roux spoke of a "frog" in her stomach. To her, the frog's loud or soft croaking and relatively abrupt movements indicated the improvement or exacerbation of the sickly phenomena. Micheline Viollet said she had "a clock at the bottom of her stomach" that indicated the present, past, and future to her. She explained that, during crises, the pendulum's movements combined with the 24 candles on both sides of her spine to enlighten her. The system, strength, and movement of the pendulum directed her predictions and guided her behavior in crisis.

Some somnambulists believe they have one or more genies or celestial spirits taking care of them. The patients talk together with their genies or spirits during crisis and often self-prescribe treatment that they no longer want—once they awaken. They ordinarily foresee this conflict between what they need to do, on the one hand, and the repugnance they feel about doing it, on the other. Writings about this tension since St. Augustine's *L'Homo duplex*[1] are extremely curious.

In any case, we find an inestimable advantage in the patients' power to predict. If asked what must be done to vanquish their repugnance so that they heal, they immediately respond, "You must do this or that." In fact, if the anticipated disgust surfaces, it is well to do what the patient says and do so immediately. He will eat without revulsion and everything will go well, although there will be no memory of the somnambulistic episode.

To many medical men, a physician's deference to his patients seems most ridiculous. Filled with principles learned at school, they feel they have the right to order patients and their pains, as Jehovah commanded the elements. Yet I must repeatedly insist (especially to young men) that physicians must defer to the patients' instinctive suggestions, especially when nerves play a significant role in the illness. Otherwise, they will never do anything good or truly useful to promote healing. In fact, without intending to, they will often oppose Nature, aggravating the illness and impeding its natural progression.

Estelle had a protecting angel (Angeline), who, though, sometimes did not speak to her. Often Estelle awaited Angeline's visit in vain. Meanwhile, we had to settle her menu and treatments for the next day. We never left the patient at day's end without going through this short planning program; it spared us many worries and questions. When Angeline did not appear, we asked Estelle about these issues. She always responded categorically. We sometimes asked her how she could tell us what she needed without Angeline. Estelle said, "It's difficult to explain that, but I see it, I feel it in myself——Wait, here is what I can tell you: I feel as though I am in a room for a great feast. I see a big table there; food garnishes it symmetrically from one end to the other. But there are a few empty places that say to me, 'Estelle? You must have this thing or that thing.' How does this happen? I have no idea, but that is as scientific as I can be."

Endnote 11

The Magnetic Formulas Patients Identify

As I wrote in Phase Two of the Magnetic Cure in Aix, Estelle had identified certain magnetic passes to put herself into crisis, get out of, stay in, or deepen the crisis state or somnambulism. She had also found passes for opening and closing her eyes at will. I call these magnetic formulae. They differ from algebraic formulae mathematicians use to resolve problems more easily and to simplify calculations. At the beginning of Phase One of the Magnetic Cure, Estelle directed her magnetic therapy. She indicated the kind of passes (magnetic formulas) that would most accelerate the progress of magnetic sleep and deepen it. In several instances, our patient demonstrated the ability to put herself into crisis, prolong this state, or, when she no longer needed it, stop the crisis. She did so at will. We see she no longer presents any nervous phenomena that belong to catalepsy. Nevertheless, she could put herself into crisis.

On the evening of August 20, 1837 (during Phase Three), Estelle sees with her eyes closed without being in the crisis state and identifies the objects around her. She cannot read, but she can see and count the lines in a given space in the first book shown to her. This event took place after several very stormy days when the patient was exposed to a highly charged electrical atmosphere that seriously tired her. Apparently this phenomenon (it has not reoccurred) was caused by this electric influence on Mademoiselle Estelle.[1]

I cannot identify the immediate cause of this phenomenon. Nevertheless, it is a new insight that one can act magnetically on oneself—in other words, one can put oneself into or out of somnambulism.

These powers, not yet studied scientifically, deserve the attention of physiologists. Such study changes—or modifies substantially—the theory of magnetism. Until now, magnetism has been regarded as an action exercised by one individual on another. This definition could not include the present phenomena. Here, magnetism is the action of one's nervous fluid on another person's fluid, or, in some cases, on one's own fluid. No magnetizer has ever acknowledged this magnetic action on oneself.[2]

The phenomenon with which Estelle so frequently presented could not be the only one of its kind. It is important that it be studied methodologically and systematically for its potential benefits for medicine, health, and treatment.

I am reminded of galvanic currents that can be established at will in an experimental closed circuit composed of a single metal (such as iron or copper) as certain parts of the circuit are momentarily heated to a higher temperature. This can be accomplished by laying an ad hoc iron circular structure over the experimental circuit. This circular structure is equipped here and there with protruding pieces of iron that can be heated intensely and readily applied to the experimental circuit.[3] Only the contact points between the experimental circuit and the ad hoc structure are heated to incandescence. This alternation between hot and cold parts of the ad hoc structure produces a galvanic current in the experimental circuit easily measurable by the magnetized needle.[4] Therefore, this metallic structure "acts on itself," even though it is composed of similar parts. To obtain the current, however, these similar parts must be put into different states with respect to one another. Otherwise, the phenomenon would not occur. Therefore, these parts, similar in their essence, become momentarily dissimilar like many disks of diverse metals in a galvanic chain.

The phenomena presented by Estelle—when she used magnetic passes on herself to become a somnambule—resemble what happens in the metallic circuit. People with nervous disorders have a hypersensitive body that can act on itself and manifest phenomena highly abnormal in all bodily functions. The body is analogous to the circuit. In the metallic circle, heat makes parts that are naturally homogenous momentarily dissimilar. Similarly, the maldistribution of animal electricity or nervous fluid causes things to happen very differently in cataleptics than in healthy people.

Philosophically speaking, nothing prevents us from thinking that the body in certain illnesses becomes a galvanic-electric apparatus. It is also true, however, that fluids follow in a particular way to maintain the healthy state. Likewise, solids have relationships of union, correspondence, and sympathy in the free, natural interplay of all healthy bodily functions. This harmony is necessarily disrupted in the abnormal or sickly state. Therefore, the innate virtual force in humans called *vis medicatrix Naturae*[5] by the ancients, must be able, according to divine design, to reestablish the equilibrium.[6]

In his *Elettrometria animale*,[7] Amoretti (1808) recognized that electricity was disseminated in the healthy, normal body in a predictable manner on the entire surface of the body. Moreover, on a given area, the electricity was in the negative state while on another area it was in the positive state. Cataleptics and magnetic somnambulists, however, feel partial sensitivity, extreme sensitivity, or no sensitivity on a given spot of their bodies. Perhaps their experience proves that this vital electricity is displaced and its natural distribution is interrupted.

These facts have been admitted by most physiologists and doctors involved with these findings. Consequently, there is sufficient reason to cite a theory already successfully applied in physics and adapt it to the human body. Although the body is composed of similar types of elements, it may—in exceptional cases—manifest a state analogous to a body composed of dissimilar parts. The possibility that man can "act magnetically" on himself (meaning everything that expression implies) does not at all contradict accepted medical principles and, even less, scientific reasoning. (To use accepted language, I use the terminology "act magnetically" rather than "magnetize himself by himself.")

If we had highly sensitive electro-metric instruments to measure the small changes of electricity in the body, we would certainly have an easy way to demonstrate the conjectures I set forth. These instruments, however, do not exist. I am hopeful that future scientific researchers will be more fortunate in this regard, and will be able to measure the movement and influence of the vital fluid in the nervous system and the entire human body.

When this measurement is available, we will be able to explain much that is still a real mystery to us. Until that time, we must consult our patients on their sensations, feelings, and experiences. Each patient can be the true judge. We must, then, listen to explanations, whatever they may be. If what the patients tell us seems believable and not too far removed from recognized truths, we must take them at their word, even if we cannot explain their accounts with our present knowledge.

I have established the possibility of nervous phenomena both through logical reasoning and through recent events involving our young patient from Neuchâtel. Long before knowing Estelle, I had observed analogous phenomena in other patients of this kind, and recently in many others. I will cite three examples, along with the formula (types of passes) each girl used. It is valuable to know that each patient varied her manner of formulating passes somewhat. Remarkably, however, each girl—unschooled in anatomy—acted on the same nerve branches to produce a particular result. Several rejected outright any suggestions I made because I would have had to obtain specific details about their inner sensations and the manner in which their crises announced themselves, formed, developed, and died out.

MICHELINE VIOLLET (SEE MY ENDNOTE 15, PATIENT 4)

When in active somnambulism, Micheline could plunge herself into syncope[g] or lethargy at will. Her formula was as follows: she would lie flat on her back in bed. Next, she crossed her right forearm on her chest and placed the tip of her middle finger of her right hand in the small indent on the neck directly above the middle of the left clavicle. Next, she would look for the corresponding spot on the other side of the neck with the middle finger of her left hand. She pressed the tip of the left middle finger on the right indent when she wanted to go into syncope.[g] A few seconds later, she obtained the desired effect.

It was understood that this state of lipothymy[g] was achieved when Micheline went limp. Lying flat on her back in bed, the parts of her body not held in place by bedding would naturally slip down. Apart from a light coloration of her face, Micheline looked exactly as if she had just died. All sensitivity to external stimuli was extinguished. Indeed, Micheline was unresponsive in all five senses. If she could hear us, she could not answer. The same was true for all regions of her body where, in all other phases of her illness, I could somehow communicate with her. This state would end spontaneously. I was not familiar with this order of phenomena at that time (1823). I knew only to observe what happened and to hope. But, I knew neither how to exploit my observations for science nor how to use them to benefit this patient or others.

ALEXANDRINE GUTTIN

Alexandrine, a young teacher in a rural community, had been suffering four to five years. She had to leave her job because of habitual palpitations and choking fits that had been attributed to an aneurysm. She came to Aix toward the end of February 1838, against the advice of her doctors and acquaintances who believed her to be incurable.

She could, at different times and with Micheline's formula, enter into syncope.[g] One day she profoundly frightened me. She was in bed where her covers happened to be piled underneath her elbows. (In this position her elbows could support her against the inevitable drop accompanying syncope.[g]) Thus, when Alexandrine used the formulas and entered the state of syncope,[g] her arms remained crossed on her chest in the syncopate position, and she could not move. She had lost her strength completely and instantaneously, as happens in self-strangulation or hanging. Her breathing was imperceptible, and her heartbeat was slowing by the second. Her feet were as cold as ice. All the vital heat seemed to be concentrated at the epigastrium[g] and heart.

Alexandrine would probably never have survived this state if I had not come to her by a chance, providential circumstance. Knowing immediately the repercussions of this position, I uncrossed her arms and placed them at her sides. After I undid the syncopate position, her arms—which had been cold as marble—warmed up. Having thus uncrossed her arms, I reawakened her vital thoracic organs by breathing air into her lungs. A few minutes later, I got unequivocal signs from Alexandrine that she heard me through the epigastrium.[g] But, she could not yet answer me. Soon, as she regained strength and life, she was able to explain all that had happened. Still in somnambulism, she furnished me with the curious remarks that follow.

I asked Alexandrine, "What happened to you?"

She replied, "I'm not sure, Doctor, but I felt myself dying."

"But how did that begin?"

"I wanted to attempt some passes (magnetic formulas), as you had told me to do. By chance, I chose the one that helped Micheline Viollet enter into syncope.[g] Feeling no pain in this state, I prolonged it a little longer than I should have. My elbows were stuck in the folds of my coverlet, so I could not move them. Suddenly, I felt like I was dying, so I wanted to move them. I could not, however, and found myself powerless."

"What did you think of then, Alexandrine?"

"Nothing, only of myself. I saw myself failing rapidly, and if you had not come to my rescue I would certainly have died."

"Have you calculated how much time you could have safely remained in that lethargic state?"

"No."

"Could you tell me approximately? One hour?"
"More."
"Two hours?"
"More."
"Six hours?"
"No, I would have succumbed before that."

Alexandrine read and wrote while in somnambulism, and she described all the phenomena of catalepsy with extraordinary intelligence. She also read the thoughts of people with her and she could see from far away. Electrification by sparks at the epigastrium,[g] sinciput,[g] and on a fairly obvious goiter at the center of her neck plunged her very promptly into magnetic sleep. I was greatly surprised by the almost complete disappearance of the goiter by means of electrification (while in crisis, she had prescribed the treatment).

Alexandrine left Aix on May 31, 1838. She was perfectly cured of her palpitations and tightness in the chest. She was feeling rather well, in general. Yet she maintained the ability to put herself into and out of crisis, open and close her eyes during a crisis, and read with her elbow, neck, and a finger on her left hand. She assured me, that, with a well-organized magnetic program, she would be completely cured of all her problems. Alexandrine wished, however, to preserve the ability to put herself into crisis with magnetism and to achieve the necessary degree of lucidity to read at will by deepening the magnetic sleep.[8]

I have shown this very remarkable example of a person's ability to act magnetically on oneself in certain circumstances. Of all the patients with this ability, Estelle used it with the most ease and speed.

Henriette Bourgeat

Henriette, from Pin, in the Canton of Virieu, continues to manifest the same phenomena. A discharge from an ordinary Leyden jar suffices to put her instantaneously into crisis and send her into lockjaw.[g] Her extremities must be relaxed with passes or magnetic formulas. Since this state is the most frequent one, I call it her ordinary crisis state. Henriette presents with syncope or apparent sleep; her face is serene. She also experiences transposition[g] of hearing to the epigastrium[g] and to the two last fingers and toes of both hands and feet. She appears to be sleeping and says nothing if no one speaks to her. Ordinarily, she lies immobile in bed, flat on her back. If she is not suffering in any other way and if someone interacts with her in this state, she responds energetically. Henriette then can identify the illnesses of those with her and prescribe appropriate treatment for them. I have seen nothing as lively as these consultations. I will describe one consultation that a patient permitted me to publish. He had told Henriette nothing of his problems and was as surprised as I was that the young girl could read his heart, mind, and body so well. Moreover, he was very happy with her prescriptions. In this state, she sometimes experiences ecstasy: she is transported to the heavens. Her face during these celestial visions becomes animated and more beautiful. She talks with the saints to whom she is particularly devoted and seems to leave this wonderful state regretfully. Upon waking, she remembers none of it.

In addition to this ordinary crisis state, Henriette is subject to crises that are much more painful. Solemn or discordant bells put her into extraordinary hysterical convulsions. She acts like carp or tench fish in a frying pan. When she came to Aix, an unexpected noise like an avalanche, the discharge of high caliber artillery, or the large drum of a military band also plunged her into this state. A cry of alarm and distress, a firebell, and the frightened cries, "Fire, fire!" unnerved her (drained all energy and strength from her arms and legs). There is also often a mild delirium, with a noticeable decrease in ordinary intellectual functions. The illness manifested in these ways before Henriette came to Aix. She still continues to suffer from them, but with infinitely less intensity. She can fall into the ordinary crisis state (not the convulsive, painful one) at will by magnetizing herself. Once in crisis, Henriette can pass from this state to the one in which she runs about at will, as Estelle did. Estelle regarded this state, as we did, as her ordinary crisis state. Throwing herself into it Estelle would say, "I am going to put myself into crisis."

The only difference between Estelle and Henriette is that Estelle put herself into the crisis state directly. Henriette achieved this same state through two steps. During the first step, she put herself to sleep by going into the ordinary crisis state. During the second step, she half-awoke from the magnetic sleep. She then had sufficient strength and nimbleness for what she needed to do. This nimbleness and strength approximated the natural ordinary state of her early childhood before she became sick.

Estelle's magnetic formulas differed a bit from Henriette's. Both formulas, however, acted on the same system of nerves (principally the fifth and seventh cervical pairs).[9] Both Estelle and Henriette came to their formulas independently without consulting each other, and with no knowledge of anatomy. Micheline Viollet and Alexandrine Guttin had also created their own formulas, simply through self-reflection and by searching for them when in trance.

On the basis of my reports and on the analogous phenomena in Estelle, Micheline, Alexandrine, and Henriette, I have reason to believe that they are probably not the only people who experience this. All somnambulists (or at least the majority of them) should be able to do as much if they concentrate on the possibility. This is what I did for Henriette, a few days before she took a trip far from Aix for more than a month. For about 30 days she would be a long way from her routine, her baths in the pools, the Scottish showers, and electricity. Henriette would be in an unknown place with many curious things: churches, public monuments, charming countryside, and lovely surroundings. Every day she would hear military music and rounds at *Champ de Mars*, bells, and drums. Her desire to make this enjoyable (but possibly risky) trip made her focus intensely on ways to put herself into crisis (this process could protect her from any dangers). I needed only two days to prepare Henriette. A great number—if not all—magnetic and spontaneous somnambulists should be able to do so as well.

My reports result from my observations and experience. I do not speak until I have observed and tested. If the circumstances are the same, I dare to guarantee that anyone who brings the same openness, patience, and tenacity to this study will learn everything I have learned.

Endnote 12

The Magnetic Power of Count Paul D.

I noted the unusual influence of Count Paul D. on Estelle and on Isaure Biron from Grenoble, another patient suffering from catalepsy. There were reciprocal influences of these patients on each other. Count Paul D., superior officer of the Russian Imperial Guard, arrived in Aix for the baths. Dr. Pizzati of Florence was no longer studying the Count, who had great magnetic power. Pizzati remembered his power well, and spoke of it with a certain satisfaction. He had sent the Count to me and seemed to relish the memory, curious about how much magnetic power the Count retained.

This military man was big and tall. Because he had the oriental habit of smoking excellent tobacco most of the day, the Count's body smelled of smoke from head to toe. He had piercing eyes, like a lynx's. He was not married and rarely went out. Instead, he spent his time reading, traveling, studying strategies, and other activities typical of the educated upper classes in Russia.

After hearing about my two patients, he expressed the desire to see them. I readily consented to his wish. I introduced them after my patients and their families agreed to the meeting. From the start, I was stupefied by the immense magnetic power he exercised on them (his gaze alone petrified Estelle). For many months, Mademoiselle Isaure's pain had abated only during catalepsy, ecstasy, and her very infrequent natural sleep—almost always filled with dreams and visions. The Count made a few calming passes (13 to 15 centimeters above her) in circles around Isaure's heart and epigastrium[g] to relieve the atrocious pain centered in the pneumogastric[g] nervous system.

Most surprising—by his will alone—Count D. could suspend my magnetic relationship already established with Isaure. Indeed, an act of will was all he needed. He could then reestablish the relationship by his own will or someone else's. Since her arrival in Aix, I had been Isaure's doctor. Her own doctor (Dr. Breton) had sent her to me, and she seemed to have full confidence in me. Count D., however, soon took over. Their connection continued until his departure from Aix.

This mutual attraction puzzled me. Its reality, however, is no less certain. I saw it repeated several times: Isaure either understood or did not understand me, depending on the Count's good pleasure. That effect could be produced only by the will of this man since he did not touch her directly or indirectly. Often I saw this phenomenon repeated. We would communicate in writing to eliminate the suspicion that the patient could understand our gestures and then bend her will to the magnetizer's. We did this even though Isaure's eyes were completely closed for the duration of her crises. Count D. also sometimes varied the phenomena exclusively at my will. He said nothing, yet he could make the patient execute orders I had communicated to him in writing exactly when I wanted, even immediately. If I suddenly decided to change my orders, Isaure executed them—not as I had first instructed—but according to my modifications.

Count D. had spoken to me of this sort of tour de force since he had done it frequently before. I was very curious to see how he would accomplish this. Admittedly, I did not believe he could do it, and I still have a great deal of difficulty believing it today, despite my patients' experiences and some magnetizers' written and spoken claims (they speak of the strength of the will that alone can produce the phenomenon). Because of my serious reservations, I gave full attention to the experiment. I observed carefully what was going on. I wanted to discover the law governing the phenomenon and draw conclusions from it that might help me accurately discern the cause of its variations in different somnambulists. Therefore, I was all eyes and ears, completely attentive to gestures, looks, and the slightest movements. Meanwhile, I admit I definitely saw Count D. destroy my relationship with my patient. I saw him repeatedly renew and suspend the therapeutic rapport[1] by a single act of his will. If not, it was caused by a kind of enchantment, if you will, produced by Count D. This enchantment, whose essence and mode were absolutely unknown to me, repeatedly operated its effect when my thought or someone else's passed through the Count's will.

This phenomenon could not be explained, as many others can be, simply by a patient reading the magnetizer's thoughts (a phenomenon for those in magnetic or spontaneous crisis much more common than usually believed). Two conditions had to occur simultaneously. Isaure had to read the thoughts of the person acting on her magnetically. Additionally, Count Paul, the magnetizer, had to will the paralysis of the organ that had established a neuropathic relationship between Isaure and me.

The explanation given for this phenomenon may not be satisfying. Nevertheless, when the Count's will had reestablished my relationship with Isaure and I asked her, "But Mademoiselle, why did you not answer me when I addressed you earlier?" she responded, "For a very simple reason, Sir. It is because you asked me nothing."

Endnote 13

Letters

From Mademoiselle Lardy, Estelle's maternal Aunt Julie, written when she returned to Neuchâtel:

September 9, 1836

Sir,

The keen interest you show our dear, young patient touches me deeply and inspires me to write what I was thinking as I neared Neuchâtel. I hear, Sir, that you are thinking of having my sister spend the winter in Aix. I am frightened, even tormented by this plan since it will be impossible to find lodging where Estelle can be sufficiently protected against the cold. At the slightest chill, the little dear's illness got worse in Peseux. We had to build double-framed windows lined with moss for her bedroom. Each day, we also had to fire up the stove three times and put kettles on to maintain the room temperature. This should come as no surprise to you, Monsieur Despine. You have seen her experience no discomfort when dressed for winter during the worst summer heat. You can also imagine the effect a draft could have on Estelle's frail, delicate body. Last summer, the first chill she experienced in Aix set back progress made by the waters. Moreover, Sir, I believe there are several months in Aix when bad weather makes using the showers and baths impossible. In addition, the carriage rides seem to do our patient good. After one more month in Aix, I think we can take advantage of the still relatively mild season to take her home for November through January. If early February appears to announce a dry, hot spring, we will prepare to return for the waters and especially for your good care.

I took the liberty to write you, Sir, for I believe my sister is not completely objective regarding Estelle. Since the death of her husband, my sister and niece have been living with us, and our hearts have adopted Estelle. In a sense, we believe we have the right to speak about Estelle's condition. I repeat, Sir, that it is impossible to find an apartment in Aix like ours here without incurring costs my sister will gladly avoid. We must also avoid boredom; it has a pernicious effect on the child. Estelle has been bored since the bad weather began. Family life will be much sweeter for her during the very dark days of winter. Indeed, her young cousins will distract Estelle. They will also amuse her—she has missed their amusements because of illness. The stay in Aix will be very difficult for my sister, who is ready to make any sacrifice. Moreover, at times Estelle needs a doctor right away. You might be in Annecy then, Sir, with your own family. Imagine my sister then, far from help and her own friends. How difficult that would be for her! On the other hand, my mother's health demands my presence here. Moreover, I am the one who helps my sister with Estelle. I could not help her in Aix. I am frightened, Sir, at the mere thought of my sister's being alone. Her health, sometimes poor, might not endure yet a third winter of hardship.

I hope you will pay me heed, Sir. I needed to share this with you. I preferred to write so that you could think about it. You need not respond. I will come to the baths for an answer.

Sincerely and with heartfelt gratitude,
Julie Lardy

A second letter from Mademoiselle Lardy following a visit to Aix after the preceding letter:

Peseux, Switzerland, December 31, 1836

Sir,

During my stay in Aix, I experienced so much happiness living near you with my sister and Estelle. I remember vividly your goodness toward us. Several times I refrained from showing you my gratitude for all of us who find in your care the attachment of a father. I have not dared take up your valuable time; you use it with a faithfulness I have often envied. Today, however, a concern overrides my discretion.

I am confused. My sister's letters give the impression that Estelle's illness has taken an uncertain turn since my departure. Headaches are violent and continual. She often has bouts of fever. I am wondering why you almost completely changed Estelle's treatments. Why do you need to magnetize her? I've heard of magnetism, but I've never seen it used. Though deeply interested in scientific advancement, Switzerland, my dear Monsieur Despine, likes the *status quo* in medicine. Therefore, new remedies are poorly understood. Indeed, I hear many conflicting opinions on magnetism. Some say, "It may do some good." Others say, "It may do some harm by agitating the patient," or "One could remain a somnambulist one's entire life!"

I am deeply convinced, Sir, that since you use magnetism, it must be necessary. I am also persuaded that you alone understand this very long, complicated, and painful illness. Therefore, I come to you to learn more about it. Please speak to me with your usual frankness.

My sister tells me that you assure her, based on what you can see, that one day she will have the joy of seeing her daughter walk. I think that perhaps you—a father who knows what it is to love a child—fear afflicting a mother. Your stoic honesty weakens when dealing with my sister. You allow her to hold onto a glimmer of hope that practically no one else in the family shares. That glimmer, alas (or rather fortunately for any mother), never completely dies out.

The frankness with which I write, Sir, must prove what you mean to me. I empty my heart to you, already feeling certain it will be heard and understood. Yes, Sir, you will understand the need I have to speak with you. There is something so clear, calm, and—at the same time—so attentive in your manner of seeing things. My most tranquil moments in Aix were those spent with you.

On the brink of a new year I cannot close my letter, my dear Monsieur Despine, without expressing my holiday wishes for you. I pray that the Lord fill you with his most abundant blessings, that He bless you with a Father's love, and that He preserve your useful, valuable life for the sake especially of your suffering patients.

Sincerely and with utmost respect and gratitude,
Julie Lardy

Response to Mademoiselle Lardy:

Aix-en-Savoie, January 10, 1837

Mademoiselle,

The letters from your sister and me should have quieted your fears about magnetism. Consequently, I will not elaborate about the marvels we have been witnessing each day since December 31. Last night when your sister was bidding her daughter goodnight at midnight, she said, "My dear Monsieur Despine, if someone told me that Estelle took down the weather vane from the church tower in Aix, I wouldn't hesitate to believe it now!"

We had just visited the Mesdemoiselles D. Many people there were playing. After the usual greetings Estelle, who herself had brought us to this home, began to tease Dr. X. about his incredulity regarding magnetism and somnambulism. She answered Dr. X.'s questions and objections with all the wisdom and intelligence of the best dialectician. Our young little patient supported her thesis with the ease of a philosopher who knows his field perfectly.

You wrote, "It is possible that Estelle will remain a somnambulist her entire life." Only God sees into the future, Mademoiselle. Yet, what if that were to happen? After all, would it not be better to see her—while in a crisis state—jump, hop around, leap over a chair, swing on a swing, climb on someone's shoulders, and do numerous antics and shenanigans than to see her crippled day and night, as she has been for more than two years? Were she to remain a somnambulist for her entire life, we would see her happy and contented, enjoying all her physical and emotional skills, and experiencing no pain or difficulties. Such an achievement would be quite wonderful for Estelle. Her family and friends would be comforted by the assurance that the paralysis was only momentary—that it could be stopped at will. They would be reassured, despite the fact that this sickly state—due neither to the compression of the spinal cord nor to its organic alteration—had destabilized the status quo in Swiss medical circles.

Your sister wrote you that Estelle did not want anyone in Neuchâtel to know she was walking. Undoubtedly, the child has reasons she does not want to tell us. She has resisted all my intercessions on behalf of her dear Aunt Julie and of Monsieur De Castella. All my own questioning has been unsuccessful. Therefore, Mademoiselle, I must insist that you say absolutely nothing about this when you write to Madame L'Hardy. Make sure no one else in your home speaks of it in letters to Aix. Otherwise, it is all right to say that you are deeply pleased to learn that Estelle's treatment seems to be progressing well, although slowly. Speak only of what Estelle writes you regarding her treatment and overall condition. Although we may say nothing to those in Neuchâtel, Estelle has allowed us to tell those in Nantes that she is beginning to walk.

Enjoy all of this in silence, Mademoiselle. Tell only Monsieur De Castella, our excellent mutual friend, of our success and miracles. Your sister and I are preparing a narrative account for you. It will be among the most curious in the medical profession. Please offer my apologies to Monsieur De Castella for not yet writing to him. Explain that I would greatly appreciate the notes he had promised me on the onset of Estelle's illness and on everything that preceded it. I need them soon to begin my case study that will figure in my next report to the government on the season of our waters.

This ailment will eventually, I hope, be healed. It still demands, however, considerable study and care. We would no doubt be unwise to interrupt its treatment. Indeed, we must avoid even the slightest interference. Moreover, Mademoiselle, fear neither for the mother nor the child. If the future is unknown except to God, it is no less true that—in all probability—Estelle's story will end well. It is not true that she has been worse since your departure than she was when you were in Aix. The disorder is better defined, that is all. I told you that, in principle, Mademoiselle Estelle's paralysis was not an ordinary paralysis. This observation has been further confirmed as the affliction has taken on a more clear-cut character.

Furthermore, my dear Lady Julie, it was not I who wanted to use magnetism. It was your sister, Estelle's mother, when she saw for herself Henriette Bourgeat's intellectual development and instinctive self-suggestions during somnambulism. As you well know, I complied with her wishes, although I had little confidence in them. Moreover, you know that Estelle vehemently opposed the idea. But our success, the unexpected and unhoped for results, was the response to our efforts! They significantly surpassed our greatest hopes and all previous treatments! Therefore, we are encouraged to continue. What do you think?

Sincerely,

Added January 11:

P.S. I was unable to post my letter yesterday. I take the opportunity, Mademoiselle, to tell you about last night's moonlit walk. We went out after 8:00 p.m. when it was −4° C. Estelle was magnetized at 5:15 p.m. I needed only 10 minutes to plunge her into the most perfect magnetic sleep. Around 6:00 p.m. she got herself dressed, as if to go visiting in the middle of the summer: a light bonnet, a veil of while tulle belonging to her mother, little slippers, a brown dress, and over it her linen shawl. (Estelle no longer tolerates silk or anything like it without fatigue. For example, she has thrown aside her little mesh linen slip.) She was armed with her mother's gold watch and mine, both hanging from her neck by a gold chain. She carried a 40 franc coin in each stocking and two big gold coins in each hand. Her eyes were "nailed shut," as she says. Estelle began to walk around the room, leaning lightly on the back of a chair. She had me follow at a distance, between 60 and 90 centimeters, without touching her. She seemed to take these precautions only for safety.

Although she seemed to be looking for something, we could not figure out what was on her mind. First, she got up on a chair; then she slipped under the table. If we asked what she wanted she would say, "Let me do as I wish, for I know what I am doing. Somnambulists do not need advice from others, especially when they don't ask for it. Do only as I say, and especially, no observations."

Finally, she found my hat! (In the apartment, she had been looking in the different places where I usually put it.) Then, overjoyed, Estelle brought it to me, put it on my head, and said, "Now, come on, Momma, get ready. Mademoiselle Amélie, quickly, quickly! Emilie, come quietly, I will lead all of you——Follow me——Although my eyes are shut tight like a box, you will see that somnambulists know very well what they are doing and where they are going——Let's go, come, follow me; today I want to spend the entire time on my feet——Momma, I don't want you to carry me for that tires and oppresses you. If I need you to, I will let you know——" After these preliminaries, Estelle took me by the hand, gave the other to her mother, and we were on our way.

The walk lasted more than two hours. She visited with her shower attendants and her little companions in misfortune still in Aix. She came to my home, leafed through two volumes of caricatures, ate some blue cheese from Sassenage and nearly a half pound of bread. She drank almost an entire glass of fresh water. Passing in front of the electric machine she said to it, "Hello, Madame Machine——not today——good-bye——good-bye——till we meet again! till we meet again!" Passing in front of the presbytery she said, "Momma, let's go see the priest. I owe him a visit——I know that he is sick." After the usual greetings, we sat down. We talked for about 20 minutes on diverse subjects, but especially about the gold watch working well, magnetism, somnambulism, and the magic influence of gold in this state. Then she said good-bye in the most gracious manner and quickly returned to St. Paul,[1] having taken this long walk on foot.

Arriving at the lodging house, Estelle also wanted to visit its proprietor, Madame Roissard. We stayed there until 11:00 p.m., chatting, laughing, kidding around, playing different games, and engaging in amusing competitions. Estelle started them and, like the most agile, clairvoyant child, was the most active participant. Meanwhile, her eyes remained completely shut.

At 11:00 p.m. she wanted to go back to her apartment. There, she continued to behave playfully, performing pranks. Next, she ate half a bowl of a celery and red carrot salad, seasoned with vinegar and very little oil. To eat this salad she used her fingers like a fork. She drank a half-cup of salad dressing and a glass of beer, with the same amount of water. A little later, she sat back up in bed and asked for magnetic passes from head to toe. Soon, her protecting spirits—Angeline, Pansia, and Zéalida—arrived. She greeted them for about 20 minutes, and discussed her diet for the next day, as well as her current medical prescriptions. She announced that in a little while (and probably much sooner than had been thought at first) they would go to the Scottish shower. She wanted to go there and come back by foot, as Henriette Bourgeat had recently been doing. It might, however, be necessary to have herself magnetized to energize her legs. This is perhaps enough for one day, isn't it, Mademoiselle Julie?

From Mademoiselle Lardy:

Peseux, Switzerland, January 21, 1837

Sir,

Your goodness and friendliness always exceed the expectations of those with whom you live and work. Your letter, dated January 10 and 11, made me feel this way very deeply. Its considerable clarity and detail about our dear little invalid raised many different questions. Consequently, I cannot refuse your invitation to see you and hear your explanations about what I do not understand.

I admit, Sir, that when I first read your letter I thought I was holding a fairy tale in my hands. Yet, I repeated to myself, "What you have read is real, and your dear little Estelle is this reality." Grave thoughts bothered me until a second, more reflective reading of the letter dissipated the effects of the first. I rejoiced in thinking that Estelle was walking. Yet, this state remains a mystery to me. First, these abilities and this movement vanish when she awakes, and then she needs gold! I don't know. I have a vague memory of paying passing attention to a conversation on galvanism. What I remember makes me believe that galvanism may relate somehow to Estelle's case. Moreover, her appetite during crises strikes me as extraordinary. She eats cheese and salad in abundance, yet she says that these hurt her greatly when she is awake. Nevertheless, Sir, it is the same body that sleeps and wakes. Likewise, the same body comes and goes so lightly dressed while sleeping, then, when awake, is paralyzed again and needs flannel, quilts, and eiderdown. Oh, if you only knew, my dear Monsieur Despine, how much I would like to be inside your head for just a few moments to get clear ideas on the causes of her behavior. I sincerely hope that, when you have some leisure, you might enlighten me and calm the chaos reigning in my mind!

There is, however, a part of the treatment in which I believe my life-long relationship and friendship with Estelle permit me to get involved. I speak of her emotional development. Do you not agree,

Sir, that it is essential never to reveal to Estelle what she says or does in the crisis state? First, the human ego loves whatever makes it stand out in a crowd. Consequently, we should distrust anything that might undermine Estelle's sense of humility. This virtue, acquired only with the help of spiritual strength from outside us, helps combat the enemies of humility within us. Moreover, our imagination is so beautiful, particularly on the rare occasions when it is focused on the noble, great, and celestial. We must question whatever draws our imagination to ourselves, principally during childhood when we are egotistical without knowing it. I entreat you, Sir, for the future of this child, to beg her mother and acquaintances never to reveal to Estelle what she says or does during the crisis state.

One thing gives me pleasure about Estelle's general state in crisis: she has the same tastes as before, when she was healthy. Did she like cheese and salad? She asks for some in the crisis state. Did she like to take walks? She takes a walk in crisis. Did she like games? She plays them in crisis. Even the stubbornness evident in her character since early childhood surfaces in her responses in the crisis state. My uncle, the minister of Colombier, is delighted by the turn Estelle's illness has taken. (He knew Dr. Pétetin and Madame Arnaud's case in Lyon.) I saw Monsieur De Castella, as you recommended. I think you will make of him a believer in magnetism. "Since you are writing to Monsieur Despine," he said, "tell him he will soon receive the notes I promised. Tell him as well that he will do me the greatest pleasure by reporting the progress of our little invalid in crisis, in her waking state (which especially interests me), the state of her natural functions, and the sensitivity of her skin, limbs, and fingertips. I wonder if they have regained their former sensitivity."

With gratitude and affection,
Julie Lardy

Response to Mademoiselle Lardy:

Aix-en-Savoie, February 6, 1837

Mademoiselle,

I am a bit late responding to your kind note dated January 21. Please forgive me. But each day I have been going from one marvel to the next. I wanted to see how long this progress would last, and it still continues!

I am very glad that Monsieur De Castella is beginning to think of magnetism as we do. Clearly, it is necessary to have a wise, critical approach in this scientific field, as in all fields where natural phenomena are not typical. It would be ridiculous, however, to reject well-observed events that others have found to be real. It would be equally ridiculous to believe everything said on this subject.

As I said above, Estelle continues to progress daily, going from one marvelous accomplishment to another. It would take volumes to describe in detail the curious scenes of somnambulism we have repeatedly witnessed. You will see the accounts in the transcription I asked your sister to write. For now, you need know only what follows.

Estelle told her mother she needed to be near me day and night. Instinctively, the child feared some annoyance that would occur and precipitate convulsions. She might even die—were I not there to provide prompt assistance. As a result, your sister had to make arrangements to change her residence. Moving Estelle from the place you stayed with her was not unwelcome. Her somnambulistic runs, that often continued into the night, had caused daily commotion. This behavior did not suit that house where, except during the season of the waters, life is very tranquil, even humdrum.

We thus returned home at the end of January. From that moment forward, we limited ourselves to diminishing Estelle's suffering, while allowing the illness to take its course. We followed it step by step, and made an effort to regularize the crises so that all of us would have less worry and aggravation.

From this moment on, we saw almost all the phenomena described by Pétetin unfold in succession. We also observed many additional phenomena that he either did not note or undoubtedly did not observe. He may have seen them as nuances of phenomena he knew and thus let them go without comment, as if they were unimportant. Everyday, Mademoiselle, we see new phenomena. They are either the type I could have predicted or the type I could not have foreseen, although analogous cases exist (I have seen them described fairly well). In this way, I can explain many things to Estelle's excellent mother. Were it not for her good sense and superior intelligence, Madame L'Hardy might fear that her sometimes ethereal Estelle might one day become crazy, dazed, or a mechanized automaton. Yet, Madame L'Hardy seems unworried when she sees my composure, the result of considerable experience with these galvanic, electric, physiological-pathological phenomena. Do not be concerned by the length or unusual nature of this note, my dear Lady Julie. Depicting such extraordinary things demands extraordinary measures, too.

It is not necessary for me to go back to the events I described to you that were repeated after a fashion almost every day. Here is what we have seen since then. Estelle and your sister settled themselves in my home, and I went on an errand in Annecy. Taking advantage of my carriage, we had made excursions in the surrounding area during Estelle's sojourn in Aix. On January 28, for example, we made a trip to Saint Innocent. Did you know, Mademoiselle, that this is one of the most beautiful sites in the Aix Basin? At that time, Estelle still had her eyes shut during the crisis state. She would come and go, examining everything with her eyes shut, just as people at twilight strain to see. Moreover, Estelle was not at all tired out by her run when she returned. On the contrary, she was very happy, still under the spell of her promenade.

Upon my return from Annecy on February 2, I found that our young patient had acquired the ability to open her eyes at will. She could also fall asleep and wake up completely at will without me, as well as come out of crisis when she wanted to. From then on, her day-to-day interactions were much easier and more pleasant. On February 3, I proposed an excursion to Saint Innocent[2] for business and then to Chambéry. Both ladies were very happy to make this trip. In this 8-hour excursion, our dear Estelle could come and go and act like an alert, healthy young woman. Moreover, people who did not know her found nothing outstanding in what she did, especially compared with others her age. They laughed and ridiculed us when we told them that Estelle was actually sleeping or that she was in magnetic somnambulism. Others, however, including Madame Roissard, Dr. Bonjean and his brother the chemist, the shower attendants and porters, and people in Saint Paul, Aix, and Chambéry could not believe their eyes when Estelle visited them on foot, straight as a soldier. Without support of any kind, she walked alone with a firm, sure step. Several of my colleagues, along with others, said out loud, "Ridiculous! Monsieur Despine has been duped! Who would have thought he would have been taken in by a child?"

Would you believe, Mademoiselle Julie, that on the day your niece made a round-trip to Saint Innocent from Aix, she also made the trip to Chambéry without once getting out of the carriage. She took a two-hour walk in Chambéry and visited several acquaintances. In freezing cold temperatures, she also went into several stores to shop on her own, not arriving home until 9:00 p.m. Would you believe that all of that happened without leaning for an instant against the carriage cushions or feeling the slightest fatigue? Since then, Mademoiselle, we have often said that we could come to surprise you in Neuchâtel! What do you think of that, Aunt Julie?

Tell our excellent friend, Dr. De Castella that, in the crisis state, her overall sensitivity is now as it was in her original state of health. Estelle's paralysis and her excessive sensitivity on the back and

Photo 13 Country home of Antoine and Péronne Despine in Saint Innocent.
Source: From the private archives of Philippe Despine, Dijon.

chest reappear after crises. When Estelle returns to her sickly state, however, she has always gained something in strength and agility and lost nothing of her other capabilities. One could say that the patient always preserves an aspect of the improvement observed during the crisis state. When considering somnambulism in and of itself, isolated from all concomitant circumstances, the state is an advantage to Estelle, although contrary to her real nature. Yesterday, Estelle spent the entire day washing a collection of fossil shells as well as cleaning out and rearranging the cabinets herself. She then took some fossil samples to her relative, Monsieur Fritz Dubois.

Do not worry about her emotional development, Mademoiselle. Somnambulists are generally inspired by the instinct of personal self-preservation. While they do what is necessary, they never go beyond it and never lose sight of the ethical principles innate in man. The sentiment of one's own dignity and refinement form the basis of their instinct. These innate principles are refined by a good education. I'll continue another time, Mademoiselle.

Best Wishes,

Excerpt from Pastor Lardy's letter to Madame L'Hardy in Aix-en-Savoie:

Colombier, Switzerland, February 13, 1837

My Dear Niece,

I have just learned that Estelle's whole story has been made public in French newspapers. I have not seen the accounts, but a friend in Neuchâtel read them and told me that Estelle was named explicitly in them.

What has happened to her is, no doubt, quite extraordinary. Yet, this story surprises me much less than it surprises others. Indeed, I have had to become involved in similar situations. I was in Geneva between 1800 and 1802 doing theology studies. In my two Physics courses, Professor Pictet was already speaking about magnetism as a kind of animal electricity. Moreover, a Monsieur Pétetin, physician in Lyon, came to the house where I lodged (in Professor De Roche's home) with a Lady Arnaud. Her son was staying in the same house. When she had been cataleptic, she engaged Dr. Pétetin's services as your daughter has engaged Dr. Despine's. Pétetin became attuned to her symptoms by chance while attending to her during the crisis state. Many things happened that—had Dr. Pétetin not seen them with his own eyes—would have seemed impossible. Indeed, much that we know today was unknown then. Pétetin published his observations in a book he brought to Professor Pictet.

Pictet described them to us in the presence of Lady Arnaud. She did not know that Pétetin was speaking about her. By then, she was completely cured. Because of my youth, I was especially impressed by these events. Since then, I met Mademoiselle Sophie R. who, soon thereafter, married Mr. Henry S. Before her marriage, she was my neighbor in Auveriner where she lived with her mother. I saw Sophie often, and she frequently spoke about Mesmer, whom she had seen in Strasbourg. She told of the extraordinary things that happened to her in the crisis state. Once, she was actually believed to be dead. Sophie had been covered in a death shroud and was going to be carried to her tomb when she came out of her 19-hour catalepsy. (In France, the dead are kept at home—and mistakenly so—for only 24 hours.) This apparent death[g] experience was repeated two or three more times. One lasted 15 hours and one, 11 hours. When she told us about them, Sophie was enjoying excellent health.

I have known and heard much more about magnetism from Madame M. de Gélieu whose proximity to Dr. Prêtre, a student of Mesmer, led her to study his doctrine. I saw her do several experiments on Charles, who was very young then. She also did several experiments on me. Our family is very electric.[3] Since my childhood, thousands of sparks come out of my hair when I comb it with a very dry comb. I would enjoy sitting in front of a mirror without a light to watch this curious spectacle. You know, my dear niece, that I am water-diviner and that more than once the diviner's rod bent in my hands when I wanted to stop it from responding. I concluded that you must not be frightened by the phenomena you witness daily and that there is reason to hope in a certain, prompt cure. Estelle is in good, capable hands with Monsieur Despine. I pray that God hear and satisfy my wishes for a complete healing soon.

Sincerely,
Pastor Lardy

Despine's note below identified the writer of the next letter.

N. B. Dr. Bottex from Lyon, a doctor at the asylum and member of several academies and learned societies, saw Estelle in Aix during the summer of 1836 and took a great interest in her. Despite the tenacity of her illness, he hoped for a cure. In conversation with Estelle's mother, Bottex predicted it would take about three years. He had also seen Mademoiselle Augustine (Picat) in fits of ecstasy and catalepsy. He was struck—as most are the first time they witness an unknown phenomenon—by the seemingly magical effect gold has on these somnambulists. He asked me to update him on Estelle's and Augustine's progress. I did that in January 1837 through a friend from Lyon (Monsieur Thiaffait) who wrote me at the beginning of February: "Your letter, my esteemed friend, surprised and pleased Dr. Bottex. He must have written you about it. He said repeatedly, 'I must leave for Aix, I want to see that.'"

From Dr. Bottex:

February 10, 1837

I read in a letter sent to me by Monsieur Thiaffait, my colleague at the *Académie de Lyon*, that you did not receive the long letter in which I wrote about various nervous maladies. I also thanked you for your effective methods and for your kind attention during my stay at your baths.

I am pleased that the interesting Mademoiselle Estelle is doing much better and that she provides you with a new, remarkable case of catalepsy and magnetic somnambulism. I hope she will still be with you by the next season of the baths and that I will have the pleasure of seeing her again there.

I spoke to you of a young man subject to very unusual hallucinations. He also presents with very serious nervous symptoms. They belong to the categories of hysteria, catalepsy, epilepsy, and mental alienation called hallucination. I told his relatives to entrust him to you. His healing—if you can manage it—will certainly not be the least interesting of your cures of nervous illnesses.[4]

I would greatly appreciate any news of Augustine Picat, our interesting patient from St. Marcellin. I hope her crises diminish sensibly—even dissipate completely.

Sincerely,

Response to Dr. Bottex:

Aix-en-Savoie, February 13, 1837

My dear and honorable colleague,

I just received your kind letter dated February 10. I wish to quickly assure you of its reception and express my gratitude. In the letter from our mutual friend, Monsieur Thiaffait, you could see I had not forgotten you, despite your long silence. I am very sorry I did not receive the long epistle you mention. Indeed, the subject about which you give me many details is of great interest to me. Therefore, I dare ask you to send it again, clearly showing Aix-en-Savoie on the address. Without this precaution, many of our letters are sent to Aix-en-Provence. The few that come back are considerably delayed.

While waiting, my dear colleague, I hurry to respond to your commission. I believe I have found what your young man needs. He must come and see what we offer. Then he will choose what suits him.

Madame L'Hardy and her dear little Estelle were very touched by their fine memories of you. Please come see them here, my dear Bottex. My letter would be too long if I wrote you everything that has happened to Estelle since this little one's time with you. This is especially true because of a fortunate coincidence (that I call providential) which led her wonderful mother to request magnetic treatments for Estelle. She simply wanted to know whether we could obtain somnambulism and whether, in this state, the patient could identify treatments more efficacious than those used thus far. Little Estelle was easily chilled; she still shivered in the middle of the summer, surrounded by comforters, cotton-wool padding, and eiderdown. She had been quickly approaching complete paralysis, having lain immobile in bed for 15 months, paralyzed from the waist down. All her doctors had treated her for spinitis, Pott's Disease.[g] This condition is characterized by a radical weakness of the bony structure of the trunk or softening of the spinal cord.

For four to five months in Aix, I attended Estelle, supervising a very methodical treatment comprised of baths, showers, massage, liniment, steam baths, brushing,[g] and electricity. Still crippled, she could barely execute small movements of extension, flexing, and abduction of her feet and toes. This little being was suffering so much; she could hold up her head only if she had very soft, cushioned feather pillows inclined at a 45° angle. I subjected this pale, sickly child to some magnetic passes on December 22, 1836 for the first time, and she became a somnambulist three days later. On December 30, she sat on her bed, and on December 31st, Estelle got up, walked alone, and offered her mother a little fruit basket that she had prepared as a New Year's gift. (Madame L'Hardy was seated at the other end of the bedroom.) She sat down on her mother's knees (something Estelle had not been able to do for more than two years) and hugged her.

The next day (January 1, 1837) she ran around the house. On January 3 she walked in the garden, dressed only in a light summer dress and a flannel shirt. She ran barefoot in the snow, saying she was not cold at all, "far from it" (it was 7–8° C above zero). On January 10, she came to see me for the first time on foot. On the 11th and the following days, she visited several people who had shown interest in her.

In all her crises until the beginning of February, she constantly had her eyes closed, "nailed shut," as she says. From then on, she found a way to open her eyes spontaneously by using her intuition during magnetic sleep. She managed this mechanically (if I can express myself in this way), by doing magnetic passes on herself and by exercising methodical, special pressure on herself with her fingers. She is led solely by instinct (she has no knowledge of anatomy) to the places on her face where the nodes or nerve ganglions connect with one another. The little invalid surely never had the slightest notion of them before her illness.

Since that time, my dear colleague, Estelle has regulated her crises in a very remarkable manner. Ordinarily, she gets up between 9:00 and 10:00 a.m. During 12 hours of active somnambulistic crisis, she acts as if she were in a perfect state of health. Anyone seeing her in this state would believe her to be well. But, when the crisis is over, Estelle is again paraplegic. She again becomes chilled, barely able to hold herself up in bed, where she is supported by cushions and pillows. Moreover, her skin becomes hypersensitive, as you witnessed in Aix. At these times, she permits only her good mother and doctor to touch her along her back and up to her ears.

Meanwhile, my dear Bottex, ever since Estelle has been able to move, her paralyzed limbs have taken on a noticeable development, and her coloring is improved. Now, instead of the pale, lifeless coloring you saw in Aix, she has a rosy, lily-white complexion. All her functions have returned to normal, as in her original healthy state. While in crisis, she eats everything with an appetite. She no longer needs laxatives or other remedies. Moreover, although our little invalid may again become paraplegic after somnambulism stops, her overall state of health always improves somewhat after each crisis. We will eventually arrive at a complete cure.

Nevertheless, my esteemed friend, if you are thinking of coming to see us during the season of the baths, do not think we will wait for you to complete the cure. In fact, if her improvement continues as it has the last two months, you may have to look for our ladies in Neuchâtel. There, your doubts—like St. Thomas's—will disappear when you witness Estelle's physical and mental abilities.

Would you believe that even several colleagues in Aix do not dare to believe their eyes? One of them recently said to my son that "Good Father Despine is letting himself be influenced by the cunning of a child." But Madame L'Hardy (who knows that this is not a case of hallucination, ambition, or a miracle) is of another opinion regarding "Good Father Despine." She is grateful and pleased to remember your consoling words last summer when you examined her sick child. And even though your predictions for a happy outcome were for the distant future, she was no less grateful that you supported her hope. You managed this, even though Estelle's nervous phenomena were serious in their essence and obscure in their nature and cause. If next month, then, you bring to Aix the patient you wrote about, you will see, Sir and well-honored colleague, that all that you perceive is not just your imagination.

You asked about Mademoiselle Augustine. Her mother sent her here a few days ago. For Augustine's benefit, I felt obliged to set three conditions on her treatment: (1) the patient will live at my home since I am too old to treat illnesses of this kind elsewhere; (2) I will be the sole person in charge of the treatment, that will continue until the patient is completely cured; (3) Augustine will choose up to three personal maids. Except for Augustine's mother, I forbid family members to be present (they might question my work). Madame Picat did not want to consent to all these conditions. She insisted that the cure be completed within three weeks, like last summer. (She had initially limited it to eight days and then to fifteen.) Madame Picat sent me some rather absurd letters on this subject and ordered Mademoiselle Josephine, Augustine's older sister who had to accompany her on the trip, to put her in the care of our friend and colleague, Dr. Vidal. He has been treating her for the last eight or ten days. She is already doing better. A brief description of her state follows.

When she left us last year,[5] Mademoiselle Augustine was in excellent health, although she still had some traces of her ecstatic state. Once home, she was sent to friends living in the country where she felt quite good and enjoyed herself during her two-week stay. Then, she was called back to St. Marcellin. This change upset her; soon the problem reappeared with its original intensity.

During the autumn and winter, Drs. Chalvet and Achard had Augustine take baths that were first cold, then lukewarm, and finally hot. These prescriptions were executed more or less regularly but without very real success. Now we will see what comes of the present treatment. We are hopeful, unless Mommy Picat decides to return her daughter to St.Marcellin on the 21st day, healed or not.[6]

Good-bye, my dear Mr. Bottex. My family, very grateful for its fond memories of you, sends wishes for your happiness and health.

Sincerely,

From Mademoiselle Lardy, written on Estelle's birthday:

Peseux, Switzerland, March 18, 1837

I was again so pleased to receive your letter and to read about Estelle, my dear Monsieur Despine. Your knowledge and experience with nervous illnesses give me great confidence in you. Thanks to you, Sir, I see obvious improvement and can hope for a cure. Yes, if it pleases God, we will see our dear little one join in the games and activities of her childhood friends.

I must tell you, Doctor, that people often speak to me about Estelle. I say as little as possible, because people's imaginations are running wild and false rumors are spreading throughout the general public. Although I am happy to see the interest people have in Estelle, I am pained to see her become an object of curiosity for some and of horror for others. Worse still, scientists debate the case and fight about this child.

I was pleased to read in my sister's letter that you were upset about a newspaper report on Estelle. We were unable to get the article, but many people were talking about it in Neuchâtel. The belief that the article came from you was surprising, since few believed my sister would have consented to it. Although I outwardly defended you saying, "No, it is not from Dr. Despine," inwardly I was fearful. Today, however, I rejoice while announcing the truth loudly!

You have undoubtedly received the long letter from Monsieur De Castella. Clearly, he is not a proponent of magnetism. He believes there has been considerable exaggeration and excitement regarding Estelle's case. I don't know what else, but it is all bad! He said, for example, "This treatment—excellent in Monsieur Despine's hands—can heal your niece. But, magnetism must not be put to general use." Therefore, we rejoice to see her improvement sustained and increased. I admit, though, that something concerns me. I have heard that Estelle has no memory of what she says or does in the crisis state. In her letters, she appears to be in her natural state. It is her! She expresses her tastes, feelings, and memories. She has kept all her faculties. What, then, is going on within Estelle? Therein lies the mystery. I must stop. Yet, the mystery is precisely what I long to understand.

A few days ago, I received a letter from Henriette Bourgeat who tells me she still has crises, but crises shorter and weaker than those she experienced in Aix. They increased following four fires that started in her village, one after the other.

My dear Sir, we do not have words to express our gratitude for all you have done for Estelle. We fear only that your goodness not exceed your strength and that your health not suffer from the exhausting care you provide. I implore you, my dear Monsieur Despine, take care of yourself and accept our warmest greetings.

Julie Lardy

To Mademoiselle Lardy, also written on Estelle's birthday:

Aix-en-Savoie, March 18, 1837

Mademoiselle,

My letter dated February 6 ended with these words, "I'll save the rest for another time." Here we are, then, at another time. But, before speaking about the marvels we have been witnessing since then, I must thank you for your March 8 letter, and tell you that we are still waiting for the notes Monsieur De Castella promised. Please tell him that we do not blame him. Moreover, he need not be concerned about compromising himself with his characteristically frank admission that he believed Estelle to suffer from Pott's Disease.[g] He is not the only one to make this assessment. I would have thought the same, had I not seen the symptoms of paralysis in cataleptics appear and disappear instantaneously under the sole influence of electricity and magnetism. I would have been among the first to believe in the powerful virtue of cauteries[g] and moxas.[g] Although I expressed doubts in the beginning, I was far from expecting the results since then.

You saw Dr. Bottex from Lyon here and know that with great interest he examined our dear Estelle. He did not—at all—expect to see her on her feet so soon. You also know that, as a doctor in the asylum and professor in the *Ecole Secondaire de Médecine* in Lyon, he was deeply involved with illnesses of the nervous system. He even wrote a well-respected short work on hallucination and abnormal sensitivity. You also know, Mademoiselle, that Bottex observed Mademoiselle Augustine Picat several times in her crises of ecstasy and catalepsy. Although he did not deny the events, he was not among their most ardent adherents. He wrote me on the first of this month, asking me about a young man who is simultaneously somnambulistic, cataleptic, and epileptic. Bottex wishes to send this patient to us. I'll copy two paragraphs from his letter:

> Let's talk about your interesting Estelle. What you tell me of her is quite remarkable. The state of somnambulism produces such extraordinary phenomena in her that you ought to publish your observations in the interest of medicine. In ancient times most events regarded as marvels were explained by the influence of animal magnetism. In Estelle's case you are like Christ and his apostles. You could almost speak their words, "Get up and walk," to your paralytics.
>
> I regret I cannot leave Lyon at this time, but I will come to see you soon. I want to witness the truly marvelous phenomena you are producing at will, so to speak, that could serve as a basis for a very interesting account on ecstasy, catalepsy, somnambulism, and animal magnetism. The instances of catalepsy and somnambulism are already so numerous that it is impossible for a judicious mind to reject them categorically. Yet, it is necessary to witness them. In this way, one can speak of the phenomena with certitude and confidence and thereby convince others of their reality. Many experiments and observations must be gathered to discover the laws governing electro-magnetic fluid. To do so in a sound manner worthy of science, only conscientious professionals with no prejudices, like you, must be involved.

You can see, Mademoiselle, that Bottex is at least half converted. Perhaps one day he will be completely convinced, like our good Dr. De Castella. The latter can speak with us frankly, without fear about his understanding of Estelle's case. He can now admit all he knows about the illness and all he thinks of it without compromising his professional integrity and his reputation.

In the past, some learned, enlightened Catholic clergy believed there was some degree of evil in magnetism. Or perhaps the problem lay in how a subject was magnetized. These religious men, charged as guardians of faith and high public morality, could never be vigilant and clever enough in this regard. Some ecclesiastics even said that the devil was mixed up in it. But, in ancient and medieval times, the greatest discoveries were treated this way. Believers in God, having little knowledge of the laws governing the universe, may mean no malice when they see the work of the devil in animal magnetism. I believe that the butchers of the Inquisition often sent innocent victims to the Devil simply for knowing a little more than they themselves knew. The Inquisitioners thought they were honoring God and serving humanity by their gravest fanaticism. They condemned to death by fire men whose talents and knowledge undoubtedly would have served humanity.

Meanwhile, the Jews in the synagogue (pundits, doctors, and wise men among them) were chosen by divine providence to preserve the books that are the foundation of faith. Consequently, these people protected and actually facilitated the dissemination of Christianity and philanthropic morality. These learned Jews, along with several thousand others, witnessed Christ's unsurpassed miracles. Yet even they attributed Jesus' miracles to Beelzebub and to Satan, despite the indisputable characteristics that proved the holy origin and veracity of His cures. Should we be surprised, then, that modern philosophers (who believe in neither God nor the Devil, who refuse to be guided by the flame of faith, and who acknowledge nothing spiritual in man) deny all that cannot be explained by the data received through the senses? They may say, "We want to see with our own eyes, touch with our own hands, direct everything ourselves!" Where would our knowledge of geography, natural history, or astronomy be, if everything had to be touched and seen and if anything not subject to such testimony were rejected?

Personally, my dear Lady Julie, I believe in God, in his marvelous, immense works, and in the laws of Nature. I include in my understanding all events that touch my senses. I also believe events perceived by others, if I am sufficiently certain that these witnesses have character, knowledge, and method that would likely preclude their being deceived and misinformed. I do not believe there is witchcraft in animal magnetism.[7] Nevertheless, this treatment must neither be used on everyone nor abandoned to just anyone. Therefore, I agree with Monsieur De Castella that magnetism must be used with discretion and discernment so that nervous impressionability, that develops from habit and its frequent reinforcement, does not degenerate into illness. The laws determining the unusual phenomena of animal magnetism are not yet sufficiently well known. Consequently, we must take one step at a time—not chancing the unknown until we are very certain of what we have seen. Any other approach would result in stupidities.

Look at Estelle! Notice that we have done nothing with this dear child for 20 days. Nothing, absolutely nothing. We simply let her act according to her instinctive impulses. That is all! I have even expressed my boredom to your sister a few times! Being with a doctor who contents himself with apparent inaction must, at times, seem ridiculous to her. She is, however, resigned to it, without remark. She has come to understand that Nature must be left on its own when Estelle is doing well. Any other mother would perhaps think I am wasting her time and money by using a passive treatment method. She knows, however, that we will act again when the improvement stops.

Estelle has just gotten over the flu. We would have killed her if we had medicated her by force. For the past three days, she's been doing quite well, and the fever during the flu had its advantages. I hope that within a month we regain all we lost and even add to what we had. Today, Estelle celebrates her twelfth birthday and begins her thirteenth year. She herself acknowledges her illness has significantly set back her education. This thought upsets her and disturbs her sense of pride. This is a good omen. Such concerns may be a great incentive for physical and mental improvement. Like you, Mademoiselle, I am confused. The facts are there, however, and we cannot turn away from the truth. We should be prudent, patient, and full of confidence in God who directs everything for the best, according to His ineffable wisdom.

You judged me well, Mademoiselle, in thinking I am not one to entertain the public with patients without their consent or that of their family. I'd be even less likely to do so by explicitly naming the patient. Moreover, the newspaper is not the best means to communicate scientific findings or to report these kinds of extraordinary events in a learned way. Each event deserves to be treated separately, in sound instructive reports. In any case, would you please send me the newspaper article in question? Its style may suggest its author to me.

Respectfully yours,

From Mademoiselle Lardy:

Peseux, April 30, 1837

Sir,

After thanking God for your successes and fine care, your spirit of observation and perseverance, the first need of my heart is to thank you. My God! If providential guidance had not led us to Aix, what would poor Estelle's life have been? Perhaps she would have been forever crippled or else burned by cauteries,[g] red-hot iron, and moxas.[g] Who in Peseux would have had the idea of administering the remedies she prescribed for herself? Her subsequent healing proved that she needed what she had prescribed. Every day in Aix, my sister and I commended each other, knowing Estelle was with you. We often admired your composure, devotion, and undaunted, persistent activity.

I wanted to bring my sister's letter to Monsieur De Castella. The facts "speak for themselves" he says. "No more objections." He is amazed, but says, "This is a unique cure and a positive result never seen before."

Monsieur De Castella said he sent you a letter he had started long ago. His son would very much like to come to Aix. He should have the time; I don't know why he doesn't go. We have several doctors in Neuchâtel who, it is said, are in a state of incurable incredulity regarding Estelle. They neither believe in nor approve of magnetism, despite its triumph over my niece's dreadful state. They say, "If she is cured, the waters are responsible."

My sister speaks to me of a return to Aix! I suppose it is still far off. If she decides to come during the season of the waters, she will catch a ride with some of the many persons your European reputation attracts to your baths. My mother—my good mother—was especially happy to read this part of my sister's letter: "So, Estelle is making progress toward a cure since a return is already being discussed!" I must tell you, dear Monsieur Despine, that since the beginning of Estelle's sickness, several among my acquaintances had come to consider her as dead. All your good news did not change their minds. A few days ago, her Godmother said, "Oh, how I pity her poor mother! She is deluding herself! Since Estelle sees into the future, the time will come when the child will predict her own death! Then, we will renounce all hope. That despair will be even worse for her poor mother!" But, ever since our dear little invalid walks, we are joyful here and share in your comforting hope.

Goodbye, Sir. I am happy to think that one day you will undoubtedly come to see us in Neuchâtel and that I will have the pleasure of showing you my devotion and gratitude in person.

Julie Lardy

From Dr. De Castella:

Neuchâtel, March 1, 1837

Sir and dear Colleague,

For a long time, I have put off sending you the details you requested on dear Estelle's illness. My work made corresponding impossible.

My goal today is to trace the principal traits of her rare, interesting condition rather than to provide a complete description. Such an account would take too long. (He continues with details already given above.)

Added April 25, 1837:

P.S. I need to ask your complete indulgence, my dear colleague, for daring to take back my letter before finally sending it out. I had the flu, and it gave me and the others no rest. Finally (and perhaps this is the primary reason), I was awaiting the outcome of Estelle's evolving symptoms! Today, there has been good, positive improvement in the patient. Thus, I have the pleasure of sending you the abridged story you requested. I ask, for my own benefit, that you tell me what you think of her present state and cure. I wish to keep your curious account. I have already done a summary of it for the *Société des Sciences Naturelles de Neuchâtel*, and I would like to complete it. This Society has taken quite an interest in the extraordinary account and would be very happy to see it among its memoirs (the first volume has just been published). Without your help, I am unable to contribute to this publication. Give my warmest greetings to our dear little patient; I am thrilled to hear of her news. Tell her she would be very kind to send me another of her "dreams."[8]

Sincerely,
De Castella

Response to Dr. De Castella:

Aix-en-Savoie, April 30, 1837

Sir and esteemed colleague,

Your letter dated March 1 and April 25 arrived only yesterday. I respond directly not to prove I am more diligent than you, but because I am less tired out from work than you are, my esteemed friend. Moreover, I wish to express how pleased I am to know that you are out of the grip of the flu and that you experienced no bad side effects. Since we had no news from Peseux, either directly or indirectly, we had reason to fear that all our friends in the Neuchâtel Basin were in trouble. But now we rest easy. I am going to reassure you about the state of our little patient who is making great strides toward health.

I would like to begin by thanking you for the very valuable documents you sent me about the early stages of Estelle's condition. A good deal of information had already been provided by Madame L'Hardy. A doctor's records, however, give details lay people cannot provide.

Indeed, a doctor often writes too much, especially in illnesses as extraordinary as Estelle's.

I have just composed a fairly detailed historical and medical account about our interesting Estelle. I proposed that my son insert it into his *Bulletin annuel des Eaux*, time permitting. It is being published now. In any case, when it is printed I will send you two copies: one for you and one for your *Société des Sciences Naturelles.* This story deserves to be preserved. I list here, for you, my esteemed colleague, what I think about Estelle and her prognosis:

- She has just resumed her baths in our establishment: she put herself into the crisis state and walked there and back. Furthermore, she began using our big pool two days ago. Already this morning she was so accustomed to this exercise that she jumped into the water like the best swimmers in Neuchâtel. She wore two life preservers, but she swam like a small fish for 30 minutes. She jumped and danced. She went the full length of the pool several times, either hopping on one foot or walking on both.
- Her excessive sensitivity on the spinal region and on the thoracic cavity is gradually diminishing. In the crisis state her hypersensitivity stops, and her body can arch backward without pain when her hands are held. At those times, her head and feet can touch on the ground!
- For a few days now, she has been walking when she is not in the crisis state. Nevertheless, her progress is slow; she takes only a few steps every evening. She now eats everything except meat. Whether in crisis or out of crisis, all her vital activities are tending to normalize. We can no longer doubt that sooner or later we will achieve a complete cure.

But, my excellent friend, I know that you will not make me state the precise moment of her cure with mathematical rigor. The patient herself, since the beginning, did not hide from us that, although things are going better by letting her continue at her own pace, we must not think she will be radically cured by spring. She said that she would still have many crises after she returns to Neuchâtel.

There, my esteemed colleague, is a rather long April Fool's surprise. We must hope that by next year Estelle will no longer send you any like it. She wanted to write you today but, instead, is participating in a donkey ride. She has not forgotten, however, the gratitude she owes you. Estelle and her mother, united in heart and sentiment, send you their friendship and affection.

To Mademoiselle Lardy:

Châtelard en Bauges, May 15, 1837

Mademoiselle,

I greatly appreciated receiving your news and the letters from Nantes. I thank you for your kind thoughts and hope I truly deserve them. My greatest strength, however, lies in letting Nature work, as long as It is not contradicting the laws of Providence. I have no merit other than applying the wisdom of the Gospel parable of the talents. Indeed, I am no more than a "good and faithful servant."[9] I am happy simply to have used this talent on behalf of your young niece. Moreover, after nearly 11 months of observation under my guidance in Aix, Estelle's excellent mother is able to direct appropriately the continuing treatment of our dear Estelle and her further instruction. Mademoiselle, even though all is going as hoped and Estelle is on her way to recovery, she is still far from being able to be assimilated into the mainstream of children her age.

You wrote about a newspaper article in which Estelle was mentioned. You promised to send it, but I have received nothing. Mademoiselle, if you cannot send it to me from Neuchâtel, would you please give me the name and date of the paper so that I can get it from Paris soon? As I told you before, this newspaper would be useful while I edit my account. Moreover, I should assess it; there may be errors in it I need to correct. You say that Estelle is interested only in herself. Someone in Marseille had questioned me about her self-absorption, so I asked Estelle. Our little imp, however, always responded, "Do you take me for a witch?" I said, "But, my dear Estelle, is it that you cannot care about others or do not want to?" She replied, "Yes, I could do it if I wanted to, but it hurts me to think of others when I have so much to think about concerning myself. You[10] and the others must leave me alone.[11] They should instead ask mercenary somnambulists who make a career of magnetic somnambulism. They will get their money's worth from them."

You know, my dear Lady Julie, that Henriette Bourgeat, Henriette Marque, Micheline Viollet, and others have also spoken to me about the same issue. If, for example, you or Estelle's sister Blanche, her mother, or grandmother were sick, I am convinced that Estelle would do all she could to answer questions concerning their health. All of you are close to her heart. For anyone not close to her, however, I can predict that it would be useless to insist on her empathy. Since her character is extremely independent, your persistence would tire her out and be harmful to her. We have seen how the slightest annoyances affect her health. My dear Lady Julie, wouldn't you agree, for example, that her treatment has recently been set back another two weeks by an incident that, for anyone else, would have had no effect?

A few days ago, Estelle lost her powerful good luck charm while she was walking in the country. She was greatly upset, and despite all that we said to reassure her (such as "money isn't everything"), she remains inconsolable. As you may well expect, her mother quickly replaced it with another coin. Estelle had told me in crisis that, although her coin was not indispensable for her cure, having it would hasten the completion of the treatment. Nevertheless, the happiness of possessing this second coin could not erase the harm done by losing the first![12]

You can see by this small example, Mademoiselle, that the most insignificant incident can slow—even halt—progress completely. Accordingly, it is so difficult to calculate accurately the length of treatment for these types of illnesses, even though things may seem to be going well.

In friendship and respect—
Your very humble and obedient servant.

Excerpt of a letter from Madame Elisa de R. to Madame L'Hardy:
Nantes, France, May 3, 1837

Dear friend,

I sent Dr. Gustave Monod from Paris some details you gave me regarding Estelle. He writes, "Obviously, there was a mistake regarding the nature of her problem. Clearly, there was no lesion of the spinal cord. Estelle seems to have been stricken with a purely neurological ailment. Her complete cure is very probable, although at some time in the future."

It appears that Dr. Monod believes these extraordinary crises should not be encouraged. He would actually prefer that they be strongly discouraged. But, is there a mother anywhere who would do such a thing?

Monod's assessment is not an official consultation, but rather the reflections of a friend. In any case, if Estelle's extraordinary state continues, my friend, I would encourage you to make very sure that magnetism (which favors and exalts these crises) is justly appropriate. At the very least, you should be assured that this treatment modality will not harm the child's future health.

Endnote 14

Patients' Weight, Skin, Hair, Etc.

A Summary of this Endnote: *Near the end of* A Curious Case of Neuropathy, *note 99 listed seven possible reasons a dissociative patient, such as Estelle, may not want her recovery to go well. Estelle claimed to be losing weight and not growing taller, yet she refused to be weighed or measured. Despine and her Mother, then, took some measurements without Estelle's knowledge, compared them privately with each other, and were gratified by the incremental growth of her limbs. Despine noted the gradual restoration of Estelle's organ systems and their functions, as he noted in other patients such as Sophie La Roche. For more than four years in Virieu, Sophie had been crippled. In Aix, people were typically drawn to her angelic sweetness and pale coloring. But when Sophie was in a crisis state, she manifested terrible anguish. Despine discreetly mentioned the development of Sophie's breasts by noting that, when she arrived in Aix, she resembled one of Raphael's virgins, even when she was lying flat on her back in bed. As her health returned, her breasts became smaller. Finally she lost most of what Despine called the rounded and very gracious forms her visitors had admired. Despine observed four skin-related phenomena in patients such as Estelle and Sophie: (1) frizzy hair and severe headaches with almost burning heat on the scalp; (2) no odor on the scalp, armpits, and feet since these patients rarely perspire; (3) unusual hairiness of the legs; (4) hardly any growth of toenails and fingernails—especially on a paralyzed limb.*

Endnote 15

Patients and Related Reports

As I mentioned in my conclusion to A Curious Case of Neuropathy, God is my witness that I have observed many times all the events involving our young Estelle. I had already seen them in many other patients stricken with illnesses of this kind. In my Introduction, I wrote that I have had the rare good fortune to encounter a fairly large number of nervous ailments among the clients at our baths. I have studied them under their different aspects and followed their progress under the influence of our balneological treatments. During most thermal seasons, my colleagues and I have observed the singular phenomena of catalepsy, variants of natural somnambulism, and many related nervous states. I now present the most remarkable cases of illnesses of this type that I know. I have organized them into three classes or series. The First Series comprises what has happened in my private practice. The Second Series involves a nomenclature of cases observed by my colleagues. The Third Series includes diverse cases not yet formally classified.

For the First Series, I selected from among my preferred patients because I could evaluate them better for having observed them myself. I recognized

- that each patient, even though afflicted with the same illness, had individual peculiarities as a function of idiosyncratic temperament, physical constitution, mood, education, and domestic habits;
- that in some patients, certain phenomena showed surprising constancy and similarity despite individual eccentricities;
- that some of these phenomena were entirely independent of the patients' will or caprice, whereas other phenomena were influenced by their will or caprice;

- that most cataleptic patients present with a particular, characteristic appearance that seems to indicate a predisposition to the disorder. This presentation is so noteworthy that a physician with a little experience of catalepsy should be able to predict a patient's degree of impressionability to magnetism. The collection of portraits[1] of more than 30 patients of this kind that I had the opportunity to see unquestionably demonstrates the truth of this principle;
- that transposition of the senses[g] always took place in my patients in whom catalepsy reached its fullest development and sleep (or the crisis state) reached its greatest depth. The external sense organs, or "organs of relationship," lost their normal impressionability to the outside world and the patients' focus turned inward.

It is important to note that even though this shift took place instantaneously, transposition of the senses[g] did not occur identically in all patients. In one patient, for example, the sense of hearing experienced the transfer; in another, the sense of sight. In some, it was the senses of taste, touch, or smell; hearing remained intact or very little altered. In a few others, transposition of the five senses[g] took place either simultaneously or successively.

These types of phenomena must be varied and unstable, because they parallel the mobility of nervous fluid, their principal instrument. To date, this nervous fluid has escaped all our investigations and all mechanical, physical, and chemical research. Although inexplicable in its nature, nervous fluid exists. It is known by its effects—always constant, identical, and regular.

There is no reason here to discuss the possibility or impossibility of transposition of the senses.[g] I leave these disputes (that are useless to science at the present time) to those who believe only what they have experimented with themselves and who espouse this well known, arrogant philosophical adage as their motto: "None but we and our friends will understand."

The Second Series involves a nomenclature of cases observed by my colleagues in Aix, or reported by scholars, doctors, or well-known foreign physicians whose authority is guaranteed.

The Third Series includes diverse cases, not yet formally classified, recorded by travelers, historians, and chroniclers. Typically, these people are not involved in healing. Yet, they are known for their truthfulness or their excellent critical approach. I do not consider all these isolated events to be catalepsy. Instead, they are nervous ailments. Their degree of exaltation, varied by circumstances, has more or less pronounced nuances of the marvelous phenomena. These phenomena characterize the physiological-pathological state known under the denominations of catalepsy, ecstasy, and somnambulism. Natural somnambulism, that manifests itself spontaneously, is similar to somnambulism. Somnambulism may manifest itself artificially and by elicitation in magnetic crisis. Although a very special state, it is only a subtype of natural somnambulism. The physiological state is the same in both cases.

I introduce these three series with the principle that everything I state regarding nervous phenomena is true, regardless of what is said. Moreover, all trustworthy doctors (such as Drs. Pétetin, Joseph Franck, Dumas, Lordat), can see these phenomena as well as I can if they want to. Their observations hold true whether or not they are initiated to physics. I argue that anyone—with my focused attention, science of criticism, and sense of careful observation—will see the same phenomena in similar circumstances.

Beginning with this principle as given and indisputable, I move on to the three series. I include patient names whenever possible (with their authorization), a summary of the phenomena, names of witnesses who can guarantee the veracity of these historical-medical reports, and the sources of the information for patients not in my own clientele (for additional information, if need be).

First Series: Patients of Dr. Despine

Patient 1: Millet, Françoise, "The Girl from Faverge"

History

At age 33, Françoise came to Aix for the waters in 1820. Her constitution was strong. Her character was good and gentle, although unusual. She lived alone in the mountains and worked as a seamstress. A catalepsy had set in when she was 23, a few weeks after a fright. No one understood what was going on. She lived alone, separate from her brothers, and often locked herself in her house for several days at a time. Since she usually attended Mass, it was often only on Sundays that people noticed her absence. Once Françoise spent an entire week in the crisis state, without eating or drinking. She was believed to be dead.[2] She continued these usual behaviors during the four years following her fright until Dr. Revil of Faverge, a learned man deeply involved in the natural sciences, was called to her village (Seytenex) for another patient in 1814. Someone spoke to him of the "dead one" (Françoise in the crisis state). Dr. Revil went to see her and quickly recognized the illness described by Pétetin, even though he had never seen it in his own medical practice. He had her brought to his office to observe her more carefully.

General Phenomena

Apparent death.[g] Insensibility. Hearing at the epigastrium[g] and the feet. Permanent cataleptic state, without any movement. Barely perceptible breathing. Upon waking, no memory of what had happened in this state, but tired and completely worn out. Fantastical visions including monsters with hideous faces, witches going to their Sabbath, souls in purgatory, and a procession of small figures carrying lighted candles.

Special Phenomena

To get Françoise out of crisis, one simply made her sit up. She would breathe, yawn, stretch out her arms, rub her eyes, and quickly say, "I want to get up." Magnetism had a great influence on this patient. Françoise would stretch her arms out toward the magnetizer's arms and follow at a distance.

Patient 2: De Roussillon, Mademoiselle Annette, from Grenoble, Born in Venice

History

Annette came to Aix in 1822 when she was 20. Smallpox occurred during her childhood. She was of a nervous constitution and experienced extreme sensitivity with physical and emotional fatigue. She had had a very good education and was very religious. Annette had been in public infrequently since the 1813–1815 political upheavals in France. Her family lost its fortune. Annette also experienced a great fright. Some nervous affectation was misdiagnosed at first, then treated as abnormal ataxic[g] fever. Her doctors did not want to classify it.

General Phenomena

MOST HIGHLY DEVELOPED phenomena of catalepsy, ecstasy, hysteria, and somnambulism. Obvious influence of animal magnetism. Noticeable advantages from treatment with the Scottish showers, electricity, rest, silence, and good music. Transposition of ALL THE SENSES.[g] Extraordinary development of intelligence. Silk, fur clothing, a magnetized bar, and Spanish wax provoked very obvious phenomena of electrical energy and its varied states (positive, negative, and combined). A lot of milk, a vegetable diet (abstinence from all kinds of meat). An impish character. Hemiplegic out of crisis. In somnambulism, she recovered her ordinary, natural state of good health.

Special Phenomena
Easy, rapid reading with her fingertips. Several letters written in somnambulism. One of them is reread and copied without any light. One of them is reread and copied whenever the light is blocked by an opaque object placed between her eyes and what she is writing. Very curious mimicking scenes. The walls of the room seem transparent, like glass; looking through them, she sees curious onlookers listening at the door. Second Sight of the Hebrides.[3] She recounts the history of her entire illness and outlines the plan to follow for healing. She adores her Father and Mother in the ordinary state, but in the crisis state they "burn" her, and she cannot tolerate them nearby.

Patient 3: Roux, Annette, from Trévignin sur Aix

History
Annette Roux came to Aix in 1822. She had bright red hair and freckles. She was quick tempered, but was repressed by her religious education. In general, she was better groomed than the usual village girl. Annette was obedient, respectful, bright, and she readily accommodated others' needs. Usually Annette was in good health. She knew how to read and write. Annette had had many frights out in the fields where she took care of the family's flock of sheep. A rural policeman turned his dog on her twice there and threatened her with a sword for resisting his offensive, inappropriate verbal advances.

General Phenomena
Her first crises were irregular but they soon normalized, reoccurring every evening at the same time. They began with the state of lethargy or apparent death.[g] Dr. Vidal immediately recognized catalepsy and the transfer of senses[g] to her epigastrium.[g] He had seen the same phenomena a few months earlier in Mademoiselle Annette De Roussillon. The illness soon developed and, little by little, all the phenomena of ecstasy, somnambulism, and catalepsy unfolded. From then on, each crisis began with an account of the episodes with the policeman and her bitter reproaches to him. Annette would swear and narrate everything the scandalous local newspapers had recounted about this pitiful character's misdeeds.

Special Phenomena
Confirmation of all the phenomena seen in the two previous cases. Use of gold to calm her pain. Singular phenomenon of hearing. Echo. Particular magnetic effects that varied depending on who administered them. Her frog[4] taught her everything. We had to wrap her head in a black coat, so that she could read with her fingertips. Musical instinct. Second Sight.[g] Galvanism is present simply by the dry contact of two or more metals. Avogadro and Michelotti's galvanic chain. Special, private language used with Micheline Viollet, another patient. The music box gave her an inexpressible pleasure, especially when placed on a hollow or resonant piece of wood furniture.

Patient 4: Viollet, Micheline, from Annecy, Seamstress

History
Viollet was in Aix beginning in 1823 at 21 years of age. She was of a sanguine,[g] lymphatic[g] constitution. Her nervous symptoms developed after some domestic problems, disagreements, and repeated frights (real or imagined). Local doctors treated Micheline unsuccessfully. One of them almost amputated her left breast that had become extremely painful and brownish in coloring, even though there was no swelling or tumescence.[g] She came to see me one day. She had heard about the cures of Françoise Millet ("The Girl from Faverge") and of Mademoiselle De Roussillon, under my care. She wished to consult me on her state and to speak of her intention to come to our baths. I took Micheline into my care; she served for a long time as one of the comparative cases for the study of cataleptic, electric, magnetic, and galvanic phenomena.

General Phenomena
We first saw all the phenomena of ecstasy and somnambulism. Those related to catalepsy did not appear until the end of the first year. Before then, these cataleptic phenomena were the only physical and emotional phenomena typical of my cataleptics that Micheline did not present. The cataleptic phenomena finally appeared the very moment a doctor from out-of-town[5] said, "If there is no catalepsy, why are you calling these patients cataleptics?" I replied, "If these phenomena characteristic of catalepsy are not yet present, they will surely appear one day."

Special Phenomena
Tremendous impressionability to magnetic force. A person to whom she was antipathetic would pass her at a distance of about 100 steps.

If she simply looked at that person, she would instantaneously be cataleptic and immobile. Micheline was seen immobilized several times in autumn, perched at the top of an ash tree where she had been gathering leaves. Galvanism was intolerably powerful. Idio-electric bodies and particularly the cat stiffened her like a statue. A spark, a spontaneous discharge of animal electric fluid, or a bright phosphorescence came out of her body without any sound. Abstinence of three months. The strongest discharges from the Leyden jar tired her less than the weakest galvanic disturbances. A music box tired her while an organ or flute gave her an indescribable pleasure. She found several magnetic formulas to alleviate her terrible headaches and cure her lockjaw.[g] Attraction and repulsion of her paralyzed body parts. Formula to identify immediately whoever was talking to her because she would often speak to one person while thinking she was speaking to another. Evocation of shadows.

Micheline Viollet and Annette Roux often served in comparative studies of the electrical value (positive or negative) of a person's nervous fluid. These are the only two cataleptics I found who were able to sympathize with one another. In the crisis state they sometimes used a language unintelligible to us yet intelligible to one another. Indeed, they understood each other very well. Silk, glass, and resin tired Micheline terribly. A running watch supported her through her lipothymies[g] and prevented a deep lethargy.

Patient 5: Marque, Mademoiselle Henriette, from Vienne in Dauphiné

History

She arrived in Aix in 1830 at the age of 21.[6] She was of a nervous constitution, and a lively, loving but jealous character. For eight years, she received a fine education at one of the best French boarding schools. Henriette experienced a great change when she left the boarding school to return home. Her parents, rich tanners who had made their fortune rapidly, were overly frugal and lived simply, even austerely. Domestic quarrels were frequent. Henriette experienced an appalling fright she hid from her family, but revealed to me in somnambulism. This terror was the determining cause of her problems. From that moment on, a gradual paralysis began on one side of her body and eventually invaded it completely. This condition gradually deprived the patient of all her social and domestic pleasures. She was obliged to leave her studies and her drawing and music, talents at which she excelled. She was melancholy and morose.

The illness was treated by the best doctors, first in Vienne and then Lyon. They diagnosed only a nervous state caused by a spinitis or irritation of the spinal cord, but without any definitive characteristics. Dr. Prunelle sent her to Dr. Vidal in Aix who soon suspected something analogous to catalepsy, an assessment I endorsed after my observation. Nothing, however, allowed for a clear diagnosis before a year and a half had passed. Finally, after five months of treatment with the waters, somnambulism (and all its related marvelous phenomena) occurred with the onset of magnetic treatment. This development happened completely by chance.

One day in mid-September 1831, I spoke to Henriette of animal magnetism. The word itself made her burst out laughing. She knew of it from public discussions of newspaper articles and gave magnetic treatment no credence. I enumerated its most remarkable phenomena: making a patient fall asleep and wake up at will by magnetic passes not involving touch; identifying one's own illness, others' illnesses, and the therapeutic means of dealing with them; seeing others' thoughts; reading any open book with the fingertips; forgetting everything upon waking. Mademoiselle Marque began to laugh and jokingly called me crazy and delusional. Ironically, she then said, "Try, we'll see."

We tried. In only five minutes, I put her into the deepest magnetic sleep. From the first day, she gave me the most curious details on the cause of her illness, its development, and the incorrect or rash use of certain medications. She spoke with confidence about the duration of her treatment, its possible cure, and its course. But she added that her cure was only tentative (although ardently desired by her and her family) since family quarrels were frequent, even constant, with all except her Father. He loved her and was excessively good but allowed his wife and children to control him. This interesting patient died in Aix in my arms in 1834, after a series of family quarrels. The last of these upset her so severely that she succumbed to its effects. Her death occurred precisely when we had some reason to hope.

General Phenomena

I was charged with Henriette's cure upon her arrival in Aix in 1831. She presented with numbness and rigidity of all body parts; habitually cold legs, as if tetanized; tongue taut and drawn to the back of her mouth; monosyllabic speech uttered slowly; writing was becoming more and more illegible, but her diction, style, and word choice had lost none of their clarity or style; vital, natural animal functions were maintained, but her legs were increasingly numb, despite the waters, steam baths, and showers. Vesicatories,[g] moxas,[g] and cauteries[g] (begun before she came to Aix and then stopped between the two treatments), antispasmodic potions, calmative drinks, and finally the treatments from 1828–1830 and the first nine months of 1831 (directed according to the most rational method) produced no improvement. If these treatments slowed the progression of the illness, their impact was not particularly noticeable. Moreover, the tetanic rigidity made her legs so icy cold that no method rectified this symptom. We observed

- passive resistance of diverse body parts that stiffened against our efforts to move them;
- extreme insensitivity of the skin from the waist to the toenails; the patient could not differentiate between cold and hot;
- contraction of the muscles of the thigh and calf creating extreme tension in her feet. Standing still was impossible since her heels could not remain flat on the floor;
- immobility in bed; we found her in the morning as she had been left the night before. If asked whether she had slept well, she responded, "No, I suffered greatly all night." We laughed, since Henriette had not moved the entire night.

On September 15, 1831, I used magnetic passes on this "living mummy" for the first time. The promptness and obvious effectiveness of animal magnetism surprised us all. From that day on, the development of her instinctive phenomena was so great and rapid that the patient herself directed her treatment. She never misdirected us. When we thought ourselves better informed (by our Hippocratic science) than the patient (by her instinctive impulse), we would change her prescriptions. We were always in error in these cases and, consequently, failed.

Special Phenomena

The trouble had started at the time of her fright. Henriette's "good sleep" every evening was nothing more than spontaneous magnetic sleep (perhaps a syncope[g]), occurring at approximately the same time as the initial fright. The intensity of the crises would increase as a result of the untimely methods that had been used to treat Henriette's extraordinary problems that may be unique in all of medicine. Likewise, the crises increased because the ongoing domestic disruptions that she internalized also heightened her emotional distress and exhausted her body. During her spontaneous magnetic sleep, she could fully appreciate exactly what was happening, without being able to remember or reveal it upon waking. She could explain it only in her magnetic sleep when she could communicate with others. We could easily, without conscious effort, regulate her crises and avoid aggravating circumstances when we could ask her, while in the crisis state, about her inner feelings and impressions. We did this by communicating with her at her epigastrium,[g] neck, or fingertips. We could then restore warmth to her hands and feet, relax her muscles, and bring back their sensitivity and movement. Her menstrual periods, that had stopped for a long time, resumed.

Everything was going well until she accidentally got an enormous burn on one knee. The incident interrupted her positive progress, reversed the effect of the medicines, decreased her natural healing instincts, and made the sickness as much humoral[g] as nervous. In such a patient, we feared the worst after the burn. When we asked her about the consequences of the burn and whether she would succumb to the considerable secretion of pus after the scabs would fall off, she responded decisively, "No. No, but it will take a long time, and I will lose a lot of my lucidity. The burn will heal perfectly, and I will have no further pain two months after the wound closes." Her prediction was completely confirmed.

The seven most remarkable phenomena the patient manifested when magnetic sleep and somnambulism were at their peak included:

- Henriette "saw" someone rummaging through her bedroom—even though she was in the baths. She told the assistants helping her with treatments and quickly sent one of them to tell this indiscrete individual not to look through her letters—at the very least.
- Her family had hired someone in 1831 to escort Henriette from Aix back to Vienne. This person, because of false or misunderstood information, accused a hostess of having stolen some items of clothing from the young patient. The hostess was indignant about the accusation and called on Mademoiselle Marque to vouch for her about the incident. Because she was in the crisis state at that moment, Henriette had not heard the incident occur. But, when the hostess placed her fingers on the patient's neck (where transposition of hearing[g] usually occurred) and asked her to reveal the truth, Mademoiselle Marque did the hostess justice and revealed several secrets regarding the old clothes in question. This scene completely mystified the person who had falsely accused the hostess.
- I interrogated Henriette about some interesting negotiations in Lyon involving an agent of mine because I had received no news. The patient explained what was happening in the negotiations each

day, point by point, and told me the chances of a favorable outcome. Finally, Henriette reported the successful conclusion of the negotiations, as well as the precise day and time of my agent's return.

- Mademoiselle Marque also "diagnosed" the type of mental alienation afflicting a certain young person thought to be mad. For a long time, this young woman had been treated unsuccessfully by Drs. Delarive and Coindet père from Geneva. Henriette identified the cause and the origin of her problems that were unknown to her doctors. She also identified the circumstances that had aggravated the illness and those that would most hinder its healing. She prescribed treatment modalities. At first, the young woman's family did not take these means seriously because they did not believe in somnambulistic phenomena. The family was very religious and educated, but not at all familiar with the physiological phenomena associated with this marvelous somnambulistic state. They were more likely to attribute these phenomena to the devil than to God. After a long time spent unsuccessfully exhausting all other resources and curative means of traditional medicine, they tried the methods suggested by our sibyl from Vienne. The young woman's family finally concurred that these methods were not contrary to religion and morality. After a course of treatment with magnetic somnambulism, the patient was completely cured. (In her hometown, the sick girl's cure was considered miraculous.) Once back in the waking state, Henriette remembered none of her prescriptions for that patient. Yet, her mere presence put Henriette into a crisis. Additionally, the two patients frightened one another and considered one another mad. This interesting patient, Henriette, identified the etiology[g] and appropriate therapies for several puzzling cases. In addition, she offered therapeutic options for cholera that coincided perfectly with what experience later proved in France to be the most appropriate ways to protect against this terrible scourge and the surest ways to heal it. For prevention, she suggested warm baths, diet, exercise, and confidence or moral strength. For healing, she suggested tonic and warm drinks, rubdowns, and steam baths for better breathing—these recommendations must be started at the onset of the illness.
- Electricity and galvanism had as much influence on this patient as on the four preceding ones. Henriette Marque proved, as Annette Roux had, the galvanic capacity of metals on Avogadro and Michelotti's scale (see my endnote 3).
- Henriette could see through opaque objects and could guess the thoughts of others. She foresaw that she would have a serious illness

that would not be fatal. This predicted time period seemed to coincide with her burn. Was the burn the "illness" she had predicted? Or, was the serious illness a humoral[g] febrile affliction that had been coming on for a long time? Or, had the burn and the abundant secretion of pus foreshadowed or perhaps even averted a more serious affliction? I cannot say for certain, but I lean more toward this latter hypothesis since it is more medical and rational.

- Henriette Marque's tutelary angel, called Emmeline, was based on a childhood friend who died when Henriette was in Aix. Henriette learned of her friend's death in a dream. Emmeline appeared to Henriette a few times in magnetic sleep and ecstasies, consoled her by sustaining her courage and hope, and signaled the most critical stages of her illness. Henriette often cried for her Emmeline. Fantastical visions, emblematic legends, word puzzles, and hieroglyphics suggested remedies and the extent of her healing. Their meaning was sometimes obvious; other times, a new somnambulistic crisis was needed for Emmeline to interpret them.

Patient 6: L'Hardy, Mademoiselle Estelle, from Neuchâtel in Switzerland

History
See A Curious Case of Neuropathy.

General Phenomena
Upon her arrival in Aix in July 1836, Estelle suffered from the most complete paraplegia, occurring after a fall on her back two years before. She had been treated without success by moxas,[g] cauteries,[g] and the application of red-hot iron. By then, she had been sickly for more than two years. Five months in Aix and a most methodical treatment with the waters in all their forms had barely brought about a slight improvement, even when aided by regularly administered electrotherapy. The patient abhorred the idea of magnetism (we used it only in the most desperate cases). Yet, it brought on the most complete somnambulism at the third session. At that point, catalepsy and ecstasy appeared and developed with rapidity, manifesting all the marvels authors have written about. From that moment on, two completely distinct states manifested themselves in our patient: the crisis state and the waking state. In the latter, Estelle was like before, chilled and paralytic. The back and spine were so sensitive that she did not want anyone to touch them. She could not leave her bed, sit at higher than a 45° angle, or deviate at all from her diet without suffering horribly.

In the crisis state, she ceased being cold; on the contrary, she had an appetite for ice baths. Snow baths restored full movement and strength to her paralyzed limbs. By the eighth magnetic session, she got up alone, dressed with dexterity, walked about, and jumped and skipped as if she were in perfect health. Momentarily, she had healthy habits and could eat everything (except meat) without problems. The sensitivity of her back had disappeared. But, as soon as crisis state stopped, she would suddenly become paralyzed and cold again. Afterward, she would have no memory of anything done or said in crisis, even though her eyes had been perfectly open.

Special Phenomena

Extraordinary development of intelligence in the crisis state. Poppy red put her into a cataleptic state as did discordant church bells. A cat "burned" her, even at a distance, and stiffened her as did silk and fur if they touched her. Transposition of all the senses[g] to different parts of the body, but most particularly to the epigastrium,[g] fingertips, wrist, and shoulder. Estelle read with her eyes more easily if she approached the book with both hands while directing her fingers on the page. Once, when she was not in crisis, Estelle was able to distinguish all the objects around her while in the most profound darkness. She prescribed for herself magnetic passes for a specific purpose. This was the first of my patients to be able to formulate magnetic passes to put herself into, to prolong, and to get out of crisis. She became a specialist in the therapeutic value of different metals. She believed the power of gold on the nervous fluid and on neuralgias to be immense. This was the first of my patients to point out a noticeable difference between the crystallized and non-crystallized form of a substance. Finally, Estelle is a new type in the order of neuralgias. Her affliction has never yet been either classified or described.

Patient 7: Voisin, Madame (Widow) Maiden Name Créput, from Lyon

History

Madame Voisin was 26 years old when she arrived in Aix in 1825. She had been a widow for three years. Her husband had died suddenly of a ruptured aortic aneurism. Since then, she was extremely sad, and she suffered continually. She was in a sickly state with the following symptoms: strong pain at her heart, palpitations, deep melancholy from the idea that she, too, was having an aneurism. In their reports, the doctors treating Madame Voisin referred to her condition as a latent pericarditis.[g] In Lyon, warm showers and steam baths (to make her sweat) always fatigued her. Lukewarm baths were more helpful since they were more temperate.

General Phenomena
Very bad headaches. She could not concentrate on anything for more than five minutes. Weakness in her legs. Her pulse had always been regular. Chronically chilled. The vegetarian diet suited her; she hated meat. Remarkable difference in sensitivity on the left and right sides from the top of her head to her feet. Her left side was more sensitive. Physiognomy of the cataleptic type. Habitual facial neuralgias alternating with heart pain. Steam baths and warm showers tired her horribly. Description of different states beginning in my endnote 4.

Special Phenomena
Everything about this state made me think that this patient owed most of her indispositions to a lack of inervation and especially to the irregular distribution of nervous fluid. Consequently, the patient was comforted when treated with Scottish Showers. When we directed a stream of cold water at the pain in the pericardial[g] region, her pain stopped instantly. Electrification by brushing[g] and sparks on the same spot dissipated the pain (however intolerable it was) and brought on sleep. I considered this tendency to sleep as a predisposition to magnetic sleep (assuming such a hypothesis to be possible).

At that time, magnetism was only a theory for me. Moreover, I admit I had no confidence in it. But, Dr. Pizzati spoke of it at length when he was in Aix. Just before he was to leave, I took advantage of this fine man's skill, by asking him to demonstrate animal magnetism on one of my patients who had never heard of this treatment but who ardently wanted to heal. Dr. Pizzati made every effort to help. Although the patient knew nothing of what we were going to do, she was completely confident and committed to the process. This was my first actual observation of and lesson on the practice of magnetism.

On August 25, 1825, Dr. Pizzati did his first magnetic test, and in less than half an hour he obtained a complete sleep. Madame Voisin experienced transposition of hearing[g] to the epigastrium,[g] and then, one after the other, all the usual phenomena of this exceptional state. Without physical contact, a relationship was established immediately between Pizzati, the magnetizer, and the patient, Madame Voisin. When the observers touched Pizzati, a relationship was established between the witnesses and Voisin. She could see only the magnetizer and those who touched him, even though there were eight or ten people in the room. She tried to walk when Pizzati commanded her to do so, but could not, so she was getting impatient. When asked why she could not walk, she responded, "I can't get up. I'm too sleepy." After her magnetic sleep, she would wake up. She would then immediately experience a natural sleep as long as her magnetic sleep. The magnetization provoked muscular contractions in the areas where passes had been done. At times Madame Voisin experienced choking, pressure on the chest, nausea, hiccoughs, belching, digestive rumbling, and other spasmodic phenomena of the organs connected to the eighth pair. All symptoms ceased under the electric influence of some sparks, an insulator stool,[7] and weak shocks.

Silk, the cat, and fur pelts did not have a very marked electric action on this patient. Madame Voisin then identified the cause, source, and necessary curative procedures of her problem. She spoke frankly about the doctors who had cared for her and insisted that cold baths

Patient 8: Ogier, Madame the Countess, from Mans, Resident of Paris

A Summary of Despine's Report on this Patient: *Because she had not been helped by doctors or various hydrotherapies, Madame the Countess Ogier, age 53, came to Aix in 1819. About 20 years earlier, she had learned—at 5:00 one morning—that her beloved daughter was dead. Every morning since then, at the same early hour, she felt a strange sensation around her heart and experienced a syncope*[g] *that lasted 15 minutes. Prone to exaggeration, she complained about weak, painful muscles in her thighs, sensitive skin, and excruciating head pain. In Aix, the warmth of the showers and the bouillon*[g] *elicited (what she called) a "choking crisis." At first Ogier could walk only 50 steps with someone's support but, after hydrotherapy, she was able to walk for 15 minutes if she used a parasol as a cane to support herself. During her second visit in 1820, Ogier walked surprisingly well, but her choking fits persisted. During two such choking crises, Despine had a direct and reasoned conversation with Ogier about her fits—afterward she did not remember that exchange. When Despine compared Ogier's syncopies*[g] *to another current patient's cataleptic states (see patient 1: Françoise Millet), he concluded from his observations that apparent lipothymy*[g] *was fundamentally a nervous state analogous to what cataleptics and somnambules present.*

and electricity were her best means of healing because (in her words) they "acted by balancing her nerves."

A Summary of Despine's Report on this Patient: *For this anxious, sanguine,*[g] *and energetic patient (25 years old), life was full of dissention and melancholy. Her complaints included cramping and spasms followed by painful muscle contractions—sometimes tetanic*[g] *or opisthotonic*[g] *or emprosthotonic.*[g] *During her crises, Berthet remained conscious. Warm hydrotherapy helped her for two or three days, but it was counterproductive if continued longer. Hydrotherapy with cool water, however, always helped. Electrotherapy stopped spasms and migrating contractions. Brushing*[g] *and sparks almost totally alleviated the tensing in her extremities. Each year for several consecutive years, she returned to Aix for treatments.*

Patient 9: Berthet, Mademoiselle, from St-Bonnet-le-château, near St. Etienne

A Summary of Despine's Report on this Patient: *Nervous and bilious by temperament, Thérèse, about 30 years old, studied and worked diligently to teach young women about moral and spiritual matters. She experienced her social position and her pedagogical responsibilities as constricting. Thérèse had been plagued by a 12-year affliction that recently assumed a particular pattern: every week, she would experience warning signs of a shift, then slip into a feverish state that lasted up to 12 hours. During this time, she would successively lose most of her senses: the attack would then climax, thereafter diminishing, as her senses would return, one after the other. She needed a day or two to recover. Despine consulted with her doctors on Thérèse's case shortly before she returned home. Her physicians called her disorder "a cerebral fever that recurred at fixed intervals." Despine, however, recognized it as a nervous affliction in the same category as catalepsy and ecstasy.*

Patient 10: Thérèse, Madame, Religious Sister from St. Pierre

A Summary of Despine's Report on this Patient: *Late last summer Augustine Rivière, age 22, was treated in Aix for 30 days. Starting at age five, she had contracted a series of illnesses. At age 15, her first periods occurred at regular intervals but varied in duration. After*

Augustine was scolded at school, she had muscle contractions and became opisthotonic.[g] *She was admitted to the Hotel Dieu in Lyon for 17 days. In Tour-du-Pin, physicians tried various interventions, but her condition deteriorated. Seven months before she came to Aix, Augustine would convulse if she were startled, emotionally stressed, or heard dissonant bells—yet she would not lose consciousness. These hysterical convulsions were characterized by a rigid twisting of her body, a stiffening of her neck, and sometimes by aphony.*[g] *Her eyes looked up and the upper eyelids came down, causing the cornea to be completely covered (now known as the Spiegel eye roll, it is used to determine a subject's hypnotizability). Augustine responded to electrotherapy in several ways: shocks from the Leyden jar restored her voice; brushing*[g] *with electrical shocks relaxed her; the application of gold relaxed her wrist to her fingers as well as her feet. When she left Aix, she was almost recovered.*

Patient 11: Rivière, Augustine, from a Small Town near Tour-du-Pin

A Summary of Despine's Report on this Patient: *Parisian newspapers had published reports on Jenny Schmitz-Baud. She worked honing components for timepieces (her father was a clockmaker). Moreover, Despine had mentioned (in his endnote 3) that Jenny, a good musician, helped during rehearsals of musical performances. Over a two-month period in 1838, this 21-year-old patient was in Aix—she felt demoralized because of many tribulations. One day she suddenly felt intense neurological pain; the next day she manifested the phenomena of ecstasy and somnambulism. Her illness included five distinct states: painful crisis, catalepsy, absorption, magnetic sleep, and active somnambulism (including ecstasy). In the way that the tutelary spirit Angeline helped Estelle, Azael guided Jenny. Several times while Jenny was mesmerized, a light or phosphorescence appeared near her as a luminous point or small globe. (Despine observed this phenomenon as a flash of light in Micheline Viollet and Annette Roux.) Jenny was very responsive to mesmerism, to timepieces, and to metals. Particularly loud sounds severely tired her; fire terrified her. Despine quoted long letters from Jenny's father and brief letters from three men who witnessed Jenny's transposition of the senses.*[g] *Jenny left Aix in mid-November. Despine wrote updates on Jenny's condition in his unpublished Travel Journal written* after *both issues of his monograph were available (A. Despine, Manuscrits).*

Patient 12: Schmitz-Baud, Madame Jenny, from Geneva

A Summary of Despine's Report on this Patient: *This 52-year-old village woman, mother of four, had been healthy most of her life. In December 1823, Jeanne Mabboux presented to Despine's father Joseph in his Annecy office (at the time, he was 88 years old) with a possible case of topical fever, a neurosis. Jeanne had complained about irregularities in her menstrual periods and their flow. She also experienced pain, particularly in her head. At first the pain came every evening. Later the discomfort came periodically: she would feel severe pain, fever, and fatigue. Customary interventions were unsuccessful. Joseph called in Antoine who detected an intense neuralgia much more functional than organic in this sanguine*[g] *woman. Jeanne knew nothing of electricity. On the basis of his experience with four cataleptics and his knowledge of Pétetin's electrotherapy, Despine applied an aigrette*[g] *and electric sparks directly to Mabboux's head. She directed Despine's applications along the serpentine path of the pain until it disappeared into the region of the big occipital indent.*[g] *These treatments were administered intermittently, and eventually Mabboux was pain free. When Mabboux's son, a philosophy student, developed headaches, Despine's electrotherapy was effective for him as well. A collateral intervention Despine used is strikingly similar to what is now called the Tapas method. Tapas discovered that having subjects hold the front and back of their heads in a specific pose could calm the body and mind.*

Patient 13: Mabboux, Jeanne from Dingi, Province of Annecy, Thône Valley

I cannot elaborate on the stories of all my patients who presented with these unusual nervous phenomena. And, although each of the numerous cases I treated was unique, all showed the same common phenomena characteristic of the singular states of catalepsy, ecstasy, and somnambulism. Therefore, I group them below by types, according to their similar presentation.

The importance of the topic and the controversial nature of my work require me to identify these patients by name. Caught between credulity on the one hand and skepticism on the other hand, we must be able, if necessary, to consult the individuals involved. I have omitted, however, the names of people who preferred to remain anonymous but authorized me to give the initials of their last names.

When catalepsy and its varieties were confused with epilepsy, witchcraft, heresy, insanity, or a spell, patients usually were considered to be incurable or were condemned to death. Doctors supporting such conclusions can be accused of tactlessness and irresponsibility. Today, however, science, humanity, compassion, and religion demand—for the general good of society—that old prejudices be questioned and that reason prevail. If, however, egotists take offense and reproach me, I will respond—with my characteristically independent spirit and with the conviction that makes martyrs—by giving this old adage...

Amicus Plato: sed magis amica veritas...[8]

Classification: Patients Grouped by Similar Cluster of Symptoms

Bedin, Mademoiselle Anna, from Lyon
Claivaz, Madame, from Martigny
Finaz, Mademoiselle Nadine, from Seyssel

Principal Phenomena Catalepsy, hysterical spasms, irregular crises of somnambulism with complete or partial transposition of the senses.[g] The illness was aggravated by warm baths. The crisis was often provoked simply by food consumption or by showering the epigastrium[g] and back with hot water at 34° C.

De St-Jean, Mademoiselle, from Lyon
Cadier, Mademoiselle, also from Lyon
Finet, Mademoiselle, also from Lyon

Principal Phenomena These three patients arrived more or less recovered at our baths. They still presented, however, with vestiges of their former nervous state such as severe headaches and an appearance typical of cataleptics.

Devinaud, Mademoiselle, from Chambéry
Thonin, Mademoiselle, from Thorens
Dunand, Mademoiselle, from Annecy
La Saizaz, from Châtelard en Bauges
Héritier, Mademoiselle, also from Châtelard-en-Bauges
Noel-Bordelin, mother, from Aix-en-Savoie

Principal Phenomena Six patients among the clientele of my father, Joseph Despine. He has written about them in detail. All showed, to a greater or lesser degree, phenomena of catalepsy, somnambulism, or ecstasy with hysterical spasms and *globus hystericus*.[g] In the days when

Joseph Despine described their condition, one did not suspect either transposition of the senses[g] or the marvels revealed both by Pétetin and magnetism.

La Roche, Sophie, from Virieu
Bourgeat, Henriette, from Pin
Guttin, Alexandrine, also from Pin
Picat, Mademoiselle Augustine, from St. Marcellin
Biron, Mademoiselle Isaure, from Grenoble

Principal Phenomena Well documented cases of catalepsy with ecstasy, syncope, somnambulism, lethargy, and transposition of the senses[g] to the epigastrium,[g] fingers and nails. The patients obey commands and often respond to the tacit will of others. Gold, with its qualities of attraction and repulsion, soothes their painful neuralgias. These patients all devise magnetic formulas to diminish, prolong, or lessen the painful crises. Idio-electric bodies "burn" them. Magnetism, electricity, and magnets act on these patients in very remarkable ways.

V. Madame, from Geneva
P. Mademoiselle Anna, also from Geneva
P. Mademoiselle Adèle, also from Geneva
P. Madame, from Grenoble
D. Madame, also from Geneva
D. Mademoiselle Elisa, from London
M. Annette, from Aix-en-Savoie
P. Mademoiselle, from la Tour-du-Pin

Principal Phenomena State of hysteria, complicated by hemiplegic or paraplegic weakness and early symptoms characteristic of catalepsy. These symptoms stop as soon as the patients are placed on the insulated stool of an electric machine. They also stop when the patients are subjected to the action of magnetism, electricity, or Scottish Showers. These patients also experience localized pain on sensitive spots, that vary in location and presentation depending on each one's circumstances. The pains characteristically occur at the top of the head or in the middle of the chest, always without loss of consciousness.

Thibaud, Mademoiselle, from Grenoble
Dumoulin, Madame, from Montfleury

Principal Phenomena Weekly cerebral fever over several years. Treated without success by quinquina[g] and other traditional measures. They were cured in Aix by a treatment that combined the baths, electricity, and quinquina.[g]

De Mont-Bellet, Monsieur, from Paris
Thomson, Monsieur, from London
Le Grand, Monsieur, from Lyon
Blanc, Monsieur Pierre, from Sallanche

Principal Phenomena Stricken with nervous symptoms resembling somnambulism and catalepsy. Marked influence of electricity and magnetic passes on these patients. These kinds of cases with their sensitivity to animal magnetism are more rare in men than in women.

Maridor, Adèle, from Neuchâtel
Grosbach, Emilie, also from Neuchâtel
Fritzché, Madame, from Vevey
Gallice, Marie, from Châtelard en Bauges
Charbonnier, Laurence, also from Châtelard en Bauges
Richard, Nanon, from Sevrier
Bocquin, Antoinette, from Aix-en-Savoie

Principal Phenomena Extraordinary sensitivity to magnetism. With magnetic passes I was able, at will, to do the following: render their skin partially or completely numb; paralyze their muscles or increase responsiveness and vital movement; provoke spasms, tetany,[g] and St. Guy's dance[g] or put the patients to sleep; cause instantaneous lockjaw[g] or suddenly transfer a spasm from one limb to another.

SECOND SERIES: PATIENTS OF OTHER DOCTORS AND REPORTS BY SCHOLARS AND SCIENTISTS

A Summary of the More Important Matters in this Series: *(1) Despine commended Dr. Pétetin's 1808 text that details 15 cases. (2) Dr. Joseph Franck collaborated with several distinguished doctors in Wilna, Lithuania to treat Louise Baerkmann who presented with catalepsy, transposition of the senses,[g] ecstasy (Baerkmann's tutelary genie, like Estelle's Angeline, collaborated in her treatment) and other phenomena. (3) In a colloquy, Dr. P. Foissac responded to Dr. Husson's* Rapport *or* Comte rendu *for the Académie Royale de Paris. In his many case reports, Husson had rasied new questions about animal magnetism. On his side, Foissac analyzed and reclassified literature on mesmerism contributed by about 30 writers. Despine was convinced that this work by Foissac was the most instructive one for the novice doctor and magnetizer. It reviews magnetic-like phenomena from ancient to contemporary times. (4) Despine stressed that, during his career, he corresponded with more than 50 colleagues—in Europe and beyond—who reported their medical cases. He confirmed that cases of magnetic somnambulism were much less rare than most persons assumed. (5) Despine acknowledged an extensive list of professionals by name and recommended two specific medical journals.*

THIRD SERIES: RELATED REPORTS OF SOMNAMBULISM AND ECSTASY

A Summary of the More Important Matters in this Series: *In his commentary here, Despine demonstrated that he understood something of the wide spectrum of dissociative phenomena as represented by the following: (1) Tarentism*[g]*; (2) the Brahmans' ecstatic state; (3) skilled Indian jugglers; (4) possession states in India; (5) some elements of witchcraft; (6) persons who could see what was happening far away (called "Second Sight"*[g] *or "Double View of the Hebrides"); (7) the experience and behavior of certain individuals such as Joan of Arc; St. Augustine; Madame De Lingré with her astounding mathematical ability; and Mademoiselle M'Evoye, as described in the Rev. Thomas Glower's account. Finally, Despine commended the editors of periodicals and newspapers who were willing to report magnetic phenomena.*

Notes

Preface and Objectives of the Author

1. In 1801, Despine earned his medical degree from Montpellier. Afterward, he took courses in medicine and surgery in Paris (one of his professors was Laennec 1781–1826). The medical schools in Paris and Montpellier espoused opposing schools of thought on mental illness. Parisian training focused on an organic, rationalistic understanding whereas Montpellier focused on a vitalistic one, believing that, in matters of health, body and spirit somehow interrelate. In Despine's day, physicians had a fairly complete awareness of the anatomy of the nervous system. The rules of physics, chemistry, mathematics, and other hard sciences influenced the mechanistic explanations physicians imposed on their understanding of the way the nervous system functions. Physicians with rationalistic, organic training believed nervous symptoms were the consequence of a malfunction of the nervous system, as a machine can malfunction. The job of the physician was to repair the "machine" through various medical interventions (such as medicines, surgeries, waters). With his vitalistic training, Despine sought other explanations and treatments for these disorders. Despine elaborates on his belief in vitalism in the text that follows.
2. These inspectors were government representatives who examined the bathing establishments and collected records.
3. In Despine's day, his town was known as Aix-en-Savoie; now the town is known as Aix-les-Bains. It will usually be called simply Aix here.
4. *Despine's footnote:* Monsieur Joseph Despine, first *Médecin Directeur* of the *Etablissement Royal des Bains d'Aix, Inspecteur des Eaux, Médecin Honoraire du Roi et de la famille royale, Proto-Médecin de la Province du Genevois, Membre de l'Académie Royale des Sciences et Arts de Turin*, died at age 95 in 1830.
5. Antoine continued his travels and his wide reading after his monograph was reissued, as shown by the entry for December 7, 1841 in his unpublished Travel Journal. The entry reported that Antoine glanced through some works on magnetism, in particular, *Sciences occultes* by Ensile Salverte and *L'Emploi thérapeutique du galvanisme* by Fabré-Palapra.

His notes on the psychofolklore publications of Théodore Bouys demonstrated his rational approach to magic and religion. Moreover, he read *Faust* by Goethe; *l'Histoire de la Révolution française* by Thiers; *La mort avant l'homme* by Roselly de Lorgues; and the *Lettres* by Madame de Sévigné. (Despine, *Manuscrits, Archives privées de Philippe Despine de Dijon*) (P. Despine, 2005).

6. *Despine's footnote:* Before the construction of the *Grand-Batîment Royal* by Victor Amé III, only 300 to 400 bathers came each year to Aix. Since that time the numbers have increased each year. In 1787, the first year that my father was called there as *Médecin-Directeur,* there were only 500. In 1790, there were 600; in 1792, there were 800. Numbers decreased under the Republic and increased again under the Consulate and the Empire. However, numbers then never fell below 1,000 and never surpassed 1,200 during even the most brilliant period of this ephemeral government. In 1816, at the time of the Restoration, there were 1,400 bathers; until 1830 numbers increased rapidly each year. Since then they have vacillated between 2,500 and 3,000 each season.
7. *Despine's footnote:* These cases were treated: complicated syphilis, white tumors, scrofulous ailments with congestion or blocking, caries and ulcers, painful tics, neuroses of all sorts, scirrhus, sores, ichthyosis, and other skin ailments.
8. After Napoleon was defeated at Waterloo, "the Treaty of Paris [1815]... ended the French Revolution and the Napoleonic era"; King Victor Emmanuel of Italy took possession of Savoie and Sardinia (Northcutt, 1996, p. 250). Therefore, the French doctor Despine practiced in Italian territory and refers to France as a "neighbor" and to French curists as "foreigners." As Mackaman (1998) explains, "Visitors to Aix-les-Bains from Chambéry—even though a mere five miles [8 kilometers] separate the two locales—were always considered to be foreigners in the eighteenth and nineteenth centuries" (p. 161).

INTRODUCTION

1. *Despine's footnote*: See these *Rapports annuels* in Chambéry at the *Bureaux de l'Intendance Générale du Duché de la Savoie*, or in Turin at the *Archives municipales des Affaires Internes.*
2. Like Mesmer, Despine believed that an imponderable magnetic nervous fluid flowed through the universe. They believed that illness was caused by the fluid's uneven distribution and flow within the patient (the fluid was attracted to or repulsed by certain areas of the body). Magnetizers used their own fluid that acted to bring about a therapeutic impact—restoring in the patient the even distribution and flow of the fluid. Despine thought this magnetism had an effect similar to the application of metals and electricity. Indeed, Mesmer actually

described himself as a "powerful animal magnet who could affect human beings" (in Fara, 2005, p. 158).

3. *Despine's footnote:* Dr. Burdin promises a prize of 3,000 francs (he has it in escrow with a Paris notary) to the person who demonstrates that he can read without using his eyes, light, or touch. By demonstrating transposition of the sense of sight in this manner, he will prove indisputably that the transposition of the senses is possible.
4. The Burdin challenge reappears later in the monograph. Despine robustly believed in the transposition of the senses—the belief that, in an altered state, a person's sense (e.g., of hearing) would be relocated from its usual anatomical place (the ears) to another part of the body (such as the feet). In Despine's time, the phenomenon was deeply controversial. The transposition of the senses figured most prominently in the case of Despine's patient Jenny Schmitz-Baud (see summary of Patient 12, Despine's endnote 15) but was mentioned as a phenomenon in several other patients.
5. A traditional Chinese medical technique, acupressure is based on the same ideas as acupuncture. Acupressure involves putting physical pressure (by hand, elbow, or with the aid of various devices) on different acupuncture points on the surface of the body. It is unclear how the Europeans in the nineteenth century acquired this skill.
6. Despine believed illness could be understood—like electricity—in terms of an excess or deficit of either positive or negative forces. In electricity, the terms positive and negative are used (respectively) for the forces of attraction and repulsion. Just as forces in nature seek equilibrium, electrotherapy or magnetism facilitated equalizing the excess of the vital fluid in the body or the deficit of it. That is, electrotherapy and magnetism restored the healthy equilibrium in a patient.
7. Artificial somnambulism (now known as heterohypnosis) is "a trance state brought about through the application of animal magnetic or hypnotic techniques" of induction (Crabtree, 1988, p. xxiii). Spontaneous somnambulism (now known as autohypnosis or self hypnosis) is "spontaneous or purposeful hypnotic trance states produced by a person within his or her own psyche. These states may include any or all of the full range of hypnotic phenomena, such as sensory alterations, anesthesia, time distortion, relaxation, age regression, and alternations in physiological functioning" (Barach, 1994, p. 11).
8. Pathological and morbid symptoms are abnormal phenomena that reflect a disease process.
9. "Although Despine was praised for being gentle with and attentive to patients, his communications with some administrators gave evidence of his confrontational side as well. In an October 15, 1834 letter to the *Intendant* Matthieux, Despine spelled out his convictions about those who exercise authority over him. 'To administer and manage the mineral waters, it is not enough simply to be a good administrator, a distinguished architect, or to be thorough and calculating, bringing

to the project the most zeal and the best will-power.... You must also be a doctor, physician, chemist...and you must also have studied in particular the application of the principles of medicine, physics, and chemistry to the science of the waters' (Archives des Thermes nationaux d'Aix-les-Bains, vol. 8). Public and private archival materials reveal an authoritative, demanding man who loved medicine but also a generous, fatherly man of the medicinal arts" (P. Despine, 2005).

10. "Jules Philippe (1883) recalled the words of Madame Necker de Saussure who had spent a long time in Despine's waiting room: 'One time, I will dress myself as a poor person so that I will have to wait less time for M[onsieur] Despine's consultation'" (1883, pp. 19–20; quoted in P. Despine, 2005).
11. *Despine's footnote:* Alexandre Bertrand, Medical Doctor, formerly a student at the *Ecole Polytechnique* and an editor for the scientific sections of *Le Globe* and *Le Temps*, died in 1831. He became famous with the publication of *Révolutions du globe*, *Lettres sur la physique*, and other classics of considerable merit. His work meets the standards of our times but remains accessible to the general public and to women. He also published on magnetism. He left his manuscripts to his two sons, both well off and destined for science. Inheriting a fortune and the talents of their father, however, did not ensure that they would inherit his taste for and interest in research. His work was to be in six big "volumes in-8." The first volume, finished just before his death, contained a curious preliminary discourse based on his experiments and tests as well as his theory concerning these unusual phenomena.
12. Ecstasy is "an elevated state of consciousness in which the subject has an awareness of spiritual things" (Crabtree, 1988, p. xxiv). The subject usually remembers the experience of ecstasy. Catalepsy is "an abnormal state characterized by a trancelike level of consciousness and postural rigidity. It occurs in hypnosis and in certain organic and psychological disorders, such as schizophrenia, epilepsy, and hysteria" (Anderson, Anderson, and Glanze, 1994, p. 277).
13. "A curist's letter (August 11, 1835) gave an unusual glimpse into two spas and reported on Alexandre Bertrand, Despine's colleague and esteemed friend. Not all visitors to the baths were completely satisfied at Despine's establishment and some complained. Such was Monsieur Francoeur, a bather at Mont d'Or, who, in the following extract from his letter to Despine, describes the practice of Dr. Bertrand who superintended the local establishment: 'Whatever care you took to establish order in your shower stalls—you have not prevented patients from waiting a long time for their turn, from fighting over their place, or from taking too long in the showers. By setting a one-hour limit for each bather, you would not only help yourself, but also your colleagues and patients. Indeed, you are

often worn out from the hard work and care you give; I heard several patients complain that they rarely see you (or the doctor they trust). Upon leaving, then, they believe they were generous to have contributed to a modest salary that they consider to be undeserved. At Mont d'Or, everything is different, and perhaps you and your colleagues should consider adopting the methods here. Each year, or more accurately in about two months, Monsieur Bertrand makes approximately 30,000 francs. True, he is the only doctor and Mont d'Or serves barely 700 patients, while your baths have over 2,000 patients. Monsieur Bertrand, however, is free to relax or make house calls to the wealthy bathers from 9 am to 2 pm and during the evening. The bathers see him two or three times each day' " (*Archives des Thermes nationaux*, vol. 10) (P. Despine, 2005).

14. Hysteria is "an emotional disturbance that manifests in a variety of physical symptoms, such as blindness, anesthesia, or paralysis. These symptoms are produced by subconscious functions that are dissociated from normal awareness" (Crabtree, 1988, p. xxiv). For some time, multiple personality disorder was classified with the hysterias, but in 1980 it was "reclassified as a dissociative disorder in *DSM-III*" (Kluft, 1984, p. 19).
15. The mobility of the nerves has nothing to do with nerves as anatomical structures. The phrase captures the temperament of the individual—more specifically, characterological qualities. Today diagnosticians would be elaborating on the character disorder, personality disorder, and Axis II phenomena associated with the individual.
16. In 1793 between August 9 and October 9, the Army of the Alps battled to take back Lyon from counterrevolutionary forces opposing the Montagnard Convention.
17. "Despine maintained a privileged relationship with the famous Professor Lordat (1775–1870) from Tourny in the Hautes-Pyrénées. Lordat taught at the *Université de Montpellier* and served as Dean of the *Département de Médecine.* During the evening of January 10, 1842, Despine 'spen[t] the evening at the house of good Papa Lordat,' his former professor. They discussed homeopathy, vitalism, and magnetism" (Despine, *Manuscrits, Archives privées de Philippe Despine de Dijon.*) (P. Despine, 2005).
18. The German scientist Alexander von Humbolt (1769–1859) coined the term "galvanism" for the electrical qualities of certain liquids and metals that, by a mutual action, generate electricity. Galvani believed that electricity was produced in the brain, conducted by the nerves, and stored in the muscles. Finger explains, "The logic was simple. Nerves are electrically excitable, and nervous energy is electrical. From these 2 premises, nervous diseases can be explained and treated as electrical breakdowns" (Finger, 2000, p. 115). Accordingly, the galvanic action of metals could restore the nervous system (pp. 111–114).

19. In the 1800s, J.C. Maxwell showed the symmetry between electricity and magnetism, and coined the term "electro-magnetic" (Robertson, 2005, p. 56).
20. "Apparent death" may be best understood as a deep trance reaction. One theory of hypnosis proposes an atavistic[g] hypothesis. In these circles, hypnosis is considered a protective defense mechanism to ward off fear and danger—based on the observation that sometimes an animal's only way to survive is to remain immobile and thus escape detection. This immobility reflex is not unlike catalepsy. In the late Enlightenment electrical apparatuses were used to resuscitate the "apparently dead," whose life signs, including pulse, indicated that the person's physical functions had stopped (Schiffer, Hollenback, and Bell, 2003, p. 156).

A Curious Case of Neuropathy with almost Complete Paralysis

1. Estelle's full name appears only once in the original French monograph. Otherwise, Despine wrote Estelle L'H. in a deliberate effort to preserve the anonymity her mother requested to protect her daughter. Estelle's full name is used here by the translators to identify the patient clearly at the beginning of the account.
2. It is Despine, rather than De Castella, who publishes this case study.
3. Estelle's mother, born in 1803 in Auvernier, Switzerland, wed François-Nicolas L'Hardy from Columbier on May 29, 1824. The couple moved to Paris where they purchased a home or land—possibly both. Estelle was born March 18, 1825. When Estelle was baptized on April 28, 1825, Charles-Louis Lardy, Deacon of Neuchâtel was her Godfather (*Archives de L'Etat,* Neuchâtel). See Despine's endnote 13 for an excerpt from Pastor Lardy's February 13, 1837 letter to his niece, Estelle's mother.
4. Estelle is actually close to ten years old.
5. Croup, an English word, describes a deep laryngeal cough traditionally associated with diphtheria. In the medical vernacular (false) croup usually means coughing due to laryngeal spasm or edema to the vocal cords.
6. She fell on November 27, 1834.
7. *Despine's footnote:* In fact, to a certain extent, this fall could well have been considered the determining cause of Pott's Disease,[g] for such a fall is the way most cases of this cruel illness begin. On the other hand, if it had been Pott's vertebral disease, it would have produced a manifest presentation of gibbosity[g] in the two years since the fall.
8. At the time, Estelle, her mother, and younger sister Blanche were living with Julie Lardy in Switzerland.

9. In ancient Rome, natural electric shocks generated by the torpedo (an electric ray fish) were used to heal headaches, gout, paralysis, and other disorders. Pieter van Musschenbroek is usually credited with the 1745 invention of the Leyden jar. This first capacitor—it accumulated and stored electrical charges—was often used in experiments and procedures (Finger, 2000, pp. 103 and 104). Since the first half of the eighteenth century, machines to generate and store electrical charges (electrotherapy) were used to heal "fevers, deafness, blindness, headaches, toothache, ulcers and sores, rheumatism, epilepsy, sciatica, cessation of menses, tapeworms, kidney stones, hemorrhoids, even severe sore throat" (Schiffer, Hollenback, and Bell, 2003, p. 137).
10. Estelle spoke of herself in the third person, for she did not say "my state of suffering" but "her state of suffering." She is experiencing depersonalization and is either subject to rapid dissociative switching that can be understood as a shift from one state of consciousness to another, from one ego state to another, or from one personality to another, or she is subject to the passive influence of other ego states on her directly.
11. *Despine's footnote:* Henriette Bourgeat, from Pin, and Mademoiselle Augustine Picat from St. Marcellin, in the *Département d'Isère.*
12. *Despine's footnote:* This is the name used in Aix-en-Savoie for what the English call a "Shower-Bath" or a "Cold Rain Bath."
13. This information is found in Despine's endnote 1 but placed here in light of its importance to the magnetic treatments.
14. One school of thought in the field of hypnosis is that the patient knows what he or she needs and that the mind, desiring healing, guides the treatment best. The viability of this belief must be questioned when working with dissociative disorders since a dissociative disorder in general and dissociative identity disorder in particular stem from trauma. Overwhelming experiences, unpredictability, and passive influence are hallmarks of these disorders; the patient feels chronically out-of-control and so wants to assert power whenever possible. The dissociative disorder therapist, therefore, must negotiate between allowing the patient some freedom to take charge and considering the possible consequences of that freedom that include destructive reenactments from the patient's past.
15. Like many who believed in the vital nervous fluid, Despine believed that its quantity and strength diminished with age.
16. *Despine's footnote:* At several places in the following account I repeat what has just been said concerning Estelle. These repetitions are necessary to convey an accurate idea of her moment-to-moment state and her rapid changes under magnetic influence.
17. In the eighteenth century, various plate machines generated electrical shocks. Glass plates fixed on an axle were rotated between stationary rubbing cushions with a crank, thereby producing electric shocks.

The number of rotations regulated the intensity and strength of the electrical charge. Generally, the machine plates had a diameter of 30–45 centimeters and included a Leyden jar, that "condensed the electric charge and enabled greater sparks to be produced" (Turner, 1998, p. 107).

18. Symptoms of lipothymy in Despine's commentary included slowed circulation, skin discoloration, dulled senses, but no loss of memory. The *Dictionnaire de médecine* (1823) referred to lipothymia as swooning or fainting or as the first stage of syncope, when the use of the senses is suspended, blood circulation is considerably slowed, and all memory is lost.
19. Despine also mentions Phase One in his endnote 3.
20. Throughout the monograph, the translators have adjusted Despine's paragraphing to bring out and clarify his units of thought. But in Despine's account of the first three magnetic sessions (December 22 through December 24) and in Madame L'Hardy's transcription of the following seven sessions (December 25 through December 31, the last one continuing into the morning of January first) the translators have strictly followed the paragraphing of the original French source. During hypnotherapeutic sessions, there is often a continuity that needs to be honored. Editorial paragraphing, then, can disrupt the natural coherence of what transpired during the magnetic (hypnotic) sessions.
21. Throws and stops are hand and arm gestures of the mesmerist that accompany the rituals of trance induction and maintenance.
22. Our two em dash punctuation (——) conveys silent pauses in Estelle's speech during magnetic sessions. These pauses typify verbalizations of people in trance and reflect the slowed speech of the hypnotic subject (and, in parallel process, the hypnotist) and the naturally slow pauses for breathing.
23. *Despine's footnote:* The actual words of the patient appear here within quotation marks. They were taken down by her mother exactly when they were said. I had engaged Madame L'Hardy to keep a record from the start of the treatment at Aix and, from the start of the magnetic sessions, to record everything noteworthy or unusual.
24. When Despine mentions his home, it is sometimes unclear whether he refers to his town home at *12 Place Carnot* near the Baths or to his country home beside Lake Bourgeat in Saint Innocent (1.5 kilometers north of Aix).
25. The translation team decided to translate some of the possessive pronouns Estelle uses when speaking of herself as "the." This choice is not a mistaken translation of the definite article used in French to speak of body parts but a deliberate attempt to convey Estelle's sense of disconnect from her own body, a way of thinking not uncommon for dissociatives.
26. When Estelle speaks of herself or her experiences, she reports them with specificity and precision. The obsessive compulsive defenses

[OC] are geared to manage the patient's anxiety. When the OC defenses fail, then the final presentation of the patient looks more like a form of hysteria. Also reported by Fink and Golinkoff (1990).

27. The word "crisis" is interchangeable with "the crisis state." Despine sometimes uses "crisis" to say that Estelle is magnetized (hypnotized); that is, she is in an altered state of consciousness. Whenever Estelle uses the French word *crise* for crisis, the English word "crisis" is used.
28. *Despine's footnote:* Yet, since the first passes of the session, Estelle had her eyes closed and had not reopened them.
29. *Despine's footnote:* Madame L'Hardy had come closer to Estelle, not knowing that, at the moment, her own "magnetic atmosphere" did not suit her daughter.
30. Here Estelle felt what Despine understood to be antipathy or revulsion toward her mother. He also discussed its counterpart—the phenomena of sympathy or attraction—a feeling of being drawn toward someone (see his endnotes 3 and 5). These terms were descriptions of electrical phenomena as well, hence the interrelatedness of electricity and nervous illness.
31. *Despine's footnote:* At this point, the patient showed the method of pressure that best suited her. Her way differed somewhat from the ordinary method magnetizers use. She wanted me to apply pressure with the ends of my thumbs against hers (by the end of the thumb I mean the pulpy part at the end of the finger) and by putting my index finger on the articulation that joins the thumb to the metacarpus.
32. In his *Apology* Socrates writes of his "daemon," a personal protecting spirit and wise guide heard only by Socrates.
33. Estelle is entirely preoccupied with herself, her experiences, and her needs. Though she acknowledges the presence of others around her, her empathy is minimal and exists primarily in the context of how others are valuable to her.
34. *Despine's footnote:* She was referring to the cemetery.
35. This is the only place the patient's full name appears in the French original.
36. Hypnosis is thought to be experienced on a continuum ranging from light trance (such as highway hypnosis) to deep trance (persons so inwardly focused that they can barely attend to external stimuli). The depth of trance indicates where the individual is on the whole continuum. A deepening of trance occurs when the hypnotized subject goes from one state of hypnotic trance to a deeper hypnotic trance. This shift can be aided by the hypnotist but is really the purview of the hypnotized subject because it is now understood that all hypnosis is self-hypnosis.
37. *Despine's footnote:* After questioning Angeline, Estelle seemed to listen for an answer, then resumed the conversation.

38. Amnesia can occur when persons are so absorbed in their inner world that they lose track of overt external reality and seem to forget (to have amnesia) for what occurred in the outside world while they were in trance.
39. The bulk of Madame L'Hardy's transcription is in the present tense in the original and in the translation. This tense powerfully conveys the simultaneity of Estelle's magnetic sessions and Madame L'Hardy's recording of them.
40. On a trip five years later, Despine again spent Christmas Day attending to patients. "In his Travel Journal, Despine recorded that, on December 25, 1841, he attended magnetic sessions at the Kunholtz household—Michel, Catherine, and Sophie. At 3:00, magnetic session at the home of Madame Médan'" (Despine, *Manuscrits, Archives Privés de Philippe Despine de Dijon.*) (P. Despine, 2005).
41. *Despine's footnote:* On her left wrist, this child Emilie was suffering from a necrotic white growth with much pus and a foul odor.
42. Another footnote below explains Despine's perception about how Emilie's social status as a servant limits her understanding of Estelle.
43. The magic lantern was a very early version of a slide projector.
44. Most landmarks Estelle mentioned appear on contemporary maps of Paris; for instance, *rue Pavée* is on the right bank of the Seine, west of the *Place de la Bastille.*
45. Estelle's phrase *la Garde* refers to the Republican Guard branch of the national police, housed (since 1819) near the *Sénat* and the *Jardin de Luxembourg* in the *rue de Tournon* (Republican, 2006).
46. Gold was known to have strong magnetic qualities, and the human body was known to be a conductor. The apparent power of gold to relax nerve tissue and to stimulate blood circulation parallels the power of electricity to influence physical ailments therapeutically.
47. For more on the impact of contacting two medals simultaneously, see Despine's endnote 3. The ancients had recognized the quality of attraction in certain elements in nature—particularly amber. When amber was rubbed, it attracted other particles to it. The word "electric" (coined in the sixteenth century by William Gilbert, 1504–1603) comes from the Greek word for "amber" (Turner, 1998, p. 107). Likewise, the Egyptians thought that loadstone—an iron ore mineral with the power of attraction—drew away pain. Volta discovered what he called "contact electricity." It resulted from the connection of two or more metals which, when rubbed, could produce electricity (Schiffer, Hollenback, and Bell, 2003, p. 124).
48. Fur generates friction electricity when rubbed against a person or rubber or some other materials that can be magnetized.
49. See Despine's endnote 3 concerning the therapeutic effects of a watch. A well running gold watch—made of magnetic gold and other metals as part of its mechanization—produces charges that stimulate

the action of the blood and muscles. A watch running slowly or not at all does not have the same therapeutic effect. (At the end of Phase Two, see comments under "Galvanism," pp. 107–08.)

50. At Dr. Despine's home on January 10, 1837, Estelle goes through two volumes of caricatures.
51. The Death of the Knights Templar refers to French knights who, in 1307, admitted to heresy in their Order—after a long imprisonment and torture by agents of King Philip IV. Even though the knights denied the heresy when they were no longer being tortured, they were killed nonetheless since earlier they had admitted heresy (Addison, 1842). Estelle likely related her own frankness during the crisis state to the Knight's pain-induced confessions.
52. *Despine's footnote:* Most of my cataleptics used this expression to indicate how tightly their eyes are closed during the crisis state.
53. *Madame L'Hardy's footnote:* Every time Estelle wants to use Monsieur Despine's watch, he must remove its cord because it is made of silk.
54. When Estelle described feeling a "burning" sensation, she sometimes developed an appetite for snow and its cooling effects—the friction electricity as well as the heat generated probably help to explain her "burning" sensations. She also speaks of Despine's fluid as "hot" and of "fire coming from his fingertips."
55. Other comments about this session appear on December 28 appear on p. 94 as Despine conclude Phase One; also see his endnote 3.
56. It is difficult to identify Madame B. She may be the machine for electrotherapy. Despine reported to Julie in a January 11 postscript (added to his letter of January 10, 1837) that, at his home the evening of January 10, Estelle had walked in front of the machine and spoke to it, calling it "Madame Machine" and said there would be no electrotherapy right now (see Despine's endnote 13). In the passage here, the conversation is about electrotherapy. Estelle had asked Angeline and Zéalida, two inner personalities or ego states, to prescribe the strength for tomorrow's electrotherapy. It is not Estelle but these two personalities who address Madame B. by name and prescribe the range to apply tomorrow. On the other hand, that Despine withholds Madame B.'s full name leads the reader to speculate that an actual Madame B. was present and that Despine decided to withhold her identity.
57. Other commentary about this session on December 29 on p. 94 as well as in endnote 3.
58. *Despine's footnote:* Here her imperious character and egotism, characteristic of somnambulists, begin to unfold.
59. Other comments about December 30 appear on p.94 in Despine's conclusion to Phase One and on p. 185 in a letter to Dr. Bottex.
60. Estelle is prescribing a homeopathic remedy.
61. The galvanic circle (circuit) is a continuous electrical current established between the two poles of a battery, an arrangement of voltaic

elements or couples with proper conductors. The conductors can be human.

62. For more on this session, see Despine's endnote 3 and his February 13, 1837 letter to Dr. Bottex in endnote 13.
63. The watch key is constructed with a flat square on its shank. Because the key is easier to grasp firmly with the flat square, it is easier to wind the watch.
64. *Despine's footnote:* Emilie, like servants throughout time, judged the methods used on Estelle according to her own limited knowledge and hearsay. She had often mocked magnetism and electricity as well as the use of snow by this pathetic person who always needed to be wrapped in eiderdown quilts. Estelle, very insulted by this, enjoyed the opportunity to scold Emilie. Estelle appeared nice as well as spiteful when she showed Emilie what she could now do. Estelle's accomplishments amazed us all. Only she looked on these newfound abilities as quite simple and natural.
65. Constant Despine, mentioned above as Antoine's firstborn, "completed his medical training in Paris and London. He practiced medicine in Turin (1830) and in Nice (1838–1844); he served as inspector of the *Thermes royaux d'Aix* (1849–53). He was made knight of the *saints Maurice et Lazare* (1853) and of *la Légion d'Honneur* (1861). His wife was Anne Soupat (they married in 1840)" (P. Despine, 2005). Constant was critical of his father's work with animal magnetism. Skeptical about its medical use, Constant was also bothered about the time his father spent with patients who often stayed in one of the two family homes.
66. *Despine's footnote:* In electromagnetic language, to "make the chain" means to link hands. The more people in the chain, the more strength Estelle claimed to receive when she was at the end of it. Every person, undoubtedly, had a part to play.
67. While mesmerized on December 30, Estelle said Despine has her permission to go to Annecy to be with family for Twelfth Night, January 5.
68. On this date Angeline first appeared.
69. Elsewhere in the monograph, Despine adds rich specificity concerning events in January during Phase One. Above all, see two letters from Despine in his endnote 13—his January 10, 1837 letter to Aunt Julie with postscript dated January 11 and his February 13, 1837 letter to Dr. Bottex.
70. An annotation above discussed sympathy and antipathy. In the crisis state, Estelle felt an affect that she could not let herself express in her usual, waking state when she showed affection for her mother. That Estelle in the crisis state has disgust for her Mother suggests that regardless of the child's stated, overt affection for her mother in Estelle's usual, waking state, she has disavowed rejection and dismell (a profoundly negative visceral response) for her mother in another

ego state. It appears that the crisis state allows Estelle to express feelings, thoughts, sensations, and behaviors that had been forbidden. The catalepsy that followed the crisis could be emblematic of the direct encapsulation of a past memory (like a blueprint) or a symbolic metaphor of a felt experience, or some combination thereof. For instance, Estelle's catalepsy could reflect her being frozen in fear at an event or frozen into silence from a family or cultural dictum or taboo. As therapy progresses and the dissociated experiences emerge more completely, Estelle will build up affect tolerance to the past experiences and will disown them less and less. Moreover, her mother will know how it feels to be rejected by Estelle (Madame L'Hardy will build up an affect tolerance to Estelle's rejection) and Estelle will "notice" that regardless of her negativity toward mother, mother does not leave.

71. A Venetian chain has interlocking segments. It is likely that, if these interlocking parts were of different metals, they may produce enough electric charge to have a healing effect.
72. *Despine's footnote:* This is undoubtedly because porcelain is only demi-vitrified.
73. Here "jump" means a substantial shift.
74. In his discussion of Phase Two, Despine refers the reader to his endnotes 4 through 12, one after another in sequence.
75. Despine's analogy lacks clarity here not because of his writing style but because his understanding of electricity was rudimentary, although typical of his time.
76. *Despine's footnote:* The word "departure" used in Chemistry and Docimacy[g] to express the separation of two metals is perhaps not, strictly speaking, the best word to describe my idea. I do not have a better word.
77. *Despine's footnote:* At present, I have Sophie La Roche, the famous miracle-worker from Dauphiné, under my care in Aix and at my home. Several newspapers wrote about her in 1834. Her affliction is very analogous to Estelle's.
78. Despine anticipates the natural progression of dissociative disorder therapy. The crisis state and the waking state are dichotomous and, at symptom onset, they emerge relatively intact and encapsulated. Or they are actually not as dichotomous as they first appear to be, and there are likely to be states that are somewhere in between. The phenomena of co-presence—when two or more personalities influence at the same time—exist with dissociative conditions. These allow for combination experiences due to the passive influence of other states on the state that is front and center. The repetitive accessing and hearing out the messages and communications of the states gradually allow the states to become less distinct and less distinguishable as they progressively blend and finally merge. Many factors influence the timeline of merging.

79. *Despine's footnote:* In other words, none of us could communicate with Estelle at sympathetic points on her skin where the transposition of her sense of hearing[g] was supposedly taking place.
80. *Despine's footnote:* These are the actual expressions the patient used. In just a few days she had consumed two liters of this liqueur.
81. *Despine's footnote:* The music box I used for my experiments played two melodies: a song of the Swiss cowherds *Le Rans-des-Vaches* and a Tyrolean[g] song.
82. *Despine's footnote:* After she got up in the morning, Estelle always put a 40 franc coin in each of her stockings. She had done so that morning. The other coins were available to her, but she took them only when she felt the need to do so.
83. The notion here is that gold had electrical qualities: it transmits energy through conductors such as the human body. Estelle had touched her mother.
84. Hallucinating is the subjective experience of perceiving what is not there (positive hallucination) such as Estelle's celestial choir or of not perceiving what is there (negative hallucination). A hallucination can be due to a psychosis or a psychotic process (the individual looses touch with reality) or a hallucination can be a phenomenon of hypnosis. When bizarre information is initially presented, the therapist may not be able to differentiate whether the patient is dealing with a psychotic or a hypnotic process. In deep trance (somnambulism), "inhibition of higher cognitive interpretations of sensory processes can occur in the form of sensory changes, visual, auditory, olfactory and gustatory hallucinations and analgesia" (Brown and Fromm, 1986, p. 8). Clinically, a therapist can interact with the patient and have an impact on a hypnotically driven hallucination and ultimately modify and eliminate it, but the therapist is less likely to have an impact on a psychotic hallucination.
85. *Despine's footnote:* On May 1, the so-called ball "burst" for the second time. From that moment on, AS ESTELLE HAD PREDICTED, she walked more easily in the somnambulistic state. Every evening, she practiced walking in the waking state and showed gradual daily improvement. This improvement was very noticeable in her walking and overall strength.
86. "Antoine's childless Uncle Jean-Baptiste Despine adopted Antoine when he was 17; not long thereafter, his uncle died. He passed his title of Baron to Antoine as well as his property—Le Châtelard—in the Bauges Mountains that stretch from Annecy to Chambéry. Antoine became known as 'The Good Lord of the Bauges' for his good works there, such as, in 1845, establishing a school in the house on his property" (P. Despine, 2005).
87. See Despine's May 15, 1837 letter to Aunt Julie written from Châtelard en Bauges in his endnote 13. He gives a somewhat different account of this lost coin episode.

88. All of Despine's dates on these pages cannot be reconciled. For instance, he gave May 26 as both the day the group left the mountains of Bauges for Annecy and the day the group returned to Aix.
89. *Despine's footnote:* This was the first time she rode a horse in this manner.
90. This child is Isaure Biron (see Despine's endnote 12).
91. Despine specifies that the weight of the crystal is two "gros." One gros equals one drachma or one-eighth of an ounce, hence the translation to "weighed almost nothing."
92. *Despine's footnote:* She showed me a crystal the length of her index finger as she described these feelings. In addition, she usually had five or six gold francs on her.
93. After this date, Despine never saw Estelle again. Ellenberger (1970) commented that "it would be interesting to know what happened to Estelle after her memorable cure" (p. 131). Philippe Despine, though, located 'two reports on Estelle L'Hardy after she left Aix-en-Savoie.' "I saw," wrote Emilie Grosbach, of Montmirail, on August 2, 1839, "Mesdames L'Hardy, and I cannot tell you how much they were asking about you, Monsieur, including the little Estelle, and how they will be happy to see you again; they expressed great pleasure at the good news that I could give them of you" (*Archives des Thermes nationaux*, vol. 13). The second report was dated more than a decade later. On January 12, 1852, Madame Liondel, governess and resident of Helsing Port, Finland, wrote to Despine: "The baroness charged me with giving you her respects and her tribute of admiration for all that Estelle [now 26 years old] had told her about you. The baroness would very much like to have the honor of meeting you" (*Archives départementales de Haute-Savoie*, 45 J 117) (Both reports quoted in P. Despine, 2005).
94. *Despine's footnote: L'Académie Royale de Médecine.*
95. *Despine's footnote:* The temperature of the lake is ordinarily between 17° and 18° C at this time of year, like all the big Alpine lakes.
96. For more on Estelle's strength, see her mother's two September 1837 letters to Despine (in Phase Three of the Magnetic Cure).
97. *Despine's footnote:* In Switzerland, *en alpage* represents time spent by the livestock in the mountains. The cattle freely graze in fresh air on immense prairies, located on the plateaus of the great Alpine chain. This is a time of great festivity. Many families live in these elevated regions for their health or for business.
98. For other commentary, see Despine's endnote 11.
99. For several reasons, the patient may not want things to go well, such as (1) however distressing the problematic symptoms are, they are familiar; (2) it is easier to keep a habit—even a bad one—than change it; (3) there may be secondary gains to the pathologic behavior; Estelle, for instance, derives much attention, special treatment, and benefits from being ill; (4) when a symptom becomes a way of life, it

begins to define the individual; therefore, getting better may feel like a loss of identity; (5) fear of failure can also interfere with recovery, in that, if patients try to get better and then do not, they must face how serious the illness or problem may be and thus need to negotiate feeling hopeless more forthrightly. In 1836, Estelle wanted to die: how would she feel if her last hope infused into Despine failed her? (6) As long as someone is ill—certainly the case with Estelle—the person is shielded from duties and responsibilities of the "mere mortal"; removing this privilege would engender stress; (7) sometimes, dissociative disorder patients have such profound self-loathing that they may sabotage efforts to get well (either consciously or not) in a very masochistic yet determined way. See Despine's endnote 14.

100. *Despine's footnote:* In Neuchâtel, the Swiss pound equals 17 ounces of Mark (or 2 Marks 1 ounce).
101. In the Notes to Despine's Introduction, see his notes 3 and 4 about the Burdin contest.
102. It is very possible that Madame L'Hardy is referring to Despine's financial difficulties. "As head of the family, Antoine sacrificed most of his fortune to revive the firm his brothers Louis and François had established at Marseille. To pay their debts, Antoine sold family possessions. His wife Péronne was worried; indeed, later the sacrifice did cause many financial difficulties. In an October 21, 1837 letter Despine wrote in Aix to Péronne in Saint Innocent, Antoine confided, 'As for your husband, my dear Péronne, finding himself now without an estate, he has nothing more to think about than science and eternity'" (*Archives départementales de Haute-Savoie, Fonds Garbillon-Despine,* 45 J 123) (P. Despine, 2005).
103. *Despine's footnote:* Estelle's story appeared in Neuchâtel in one of these newspapers: *Le Temps*, *Le Globe*, or *Le Journal du commerce.* Perhaps it appeared in all three. I am neither aware of the author nor can I verify the accuracy of the accounts. Despite considerable effort and research, I have been unable to find these articles.
104. *Despine's footnote:* In the canton of Neuchâtel, this educational establishment is famous for its religious principles and for its excellent education of young ladies.
105. *Despine's footnote:* These patients are Sophie La Roche from Virieu, Henriette Bourgeat from Pin, and Alexandrine Guttin from the same canton of Virieu in the *Département d'Isère.*

Endnote 3: Electric and Galvanic Action of Metals

1. Within 3 months, Estelle would turn 12.
2. *Despine's footnote:* This big coin was called a "carlino," after Charles-Emmanuel II of Savoy who had it minted when he established a

new monetary system. Since the gold was worth more than the face value of the coins, however, these pieces promptly disappeared from circulation. No more of them could be found except as a type of investment in wealthy families. A family friend had lent Annette a carlino.

3. *Despine's footnote:* According to Avogadro and Michelotti (1823, *Annals de physique et de chimie, T. XII, avril*), the sequence from the extreme negative to the extreme positive is platinum, gold, silver, mercury, arsenic, antimony, cobalt, nickel, copper, bismuth, iron, tin, lead, and zinc.
4. The so-called magic tile, invented in 1747 by Englishman Doctor Bevis, was made of a square piece of glass covered with a thin sheet of tin. It produced the same electrical effect as the Leyden Jar, invented a bit earlier (Magic Tile).
5. *Despine's footnote:* Later, I had four-sided disks made whose two opposing angles served as axes to set them in motion. This way, there was no need for the steel points. Then, my experiments were more positive. They are the foundation of the theory on which my monograph is based. All my patients put the purest gold at the extreme negative of the series of metals. They often placed platinum between gold and silver. In crisis, however, they always preferred gold. It lessened the neurological pain tiring them. Since 1823, I have collected much information on gold.
6. Manheim gold is composed of three parts copper and one part zinc (Moore, 1873, p. 155).
7. This doctor and patient are mentioned in the summary of the Second Series, Despine's endnote 15.

Endnote 4: Hysteria and Its Diverse Phenomena

1. In Despine's time, scientists already knew that the nervous system was divided into the activating sympathetic nervous system and the agonist parasympathetic nervous system.
2. "Sympathetic neurosis" referred to a problem in the normal functioning of the sympathetic nervous system.
3. François-Joseph-Victor Broussais (1772–1838), professor of medicine in Paris, strongly promoted the depletive procedures of applying leeches and using bloodletting. During the 1832 cholera epidemic (when Estelle's father died) these procedures weakened many victims.
4. For treatment with the electric bath, patients sat in an insulated chair or lay in an insulated bed while holding a metal wire connected to a machine's prime conductor. The electric effect was sustained on the

periphery of the body. Lucas explains, "We know the subject had the sensation of being enveloped in a spider-web" (Lucas, 1900, p. 161).

Endnote 5: Imitation, Echo, Attraction-Repulsion

1. See Paul Moreau's (1804–1884) record of what Doctor Boerhaave witnessed at this school located a few kilometers from Leyden in Holland in *De la folie des enfants,* pp. 380–398.
2. Estelle witnessed Henriette Bourgeat's episode, as briefly mentioned in Treatment, the First Stage and Second Stage.

Endnote 6: Various States of Somnambulism

1. Somatic memory is the body memory of an experience. Either it occurred in reality or was experienced as if it occurred. A somatic memory has taken a physical form rather than an emotional form. Current traumatologists believe that the affect the mind cannot tolerate gets stored in the body. The spasm and catelepsy Despine mentioned are two of many possible presentations of what today is considered somatic memory.

Endnote 8: My Professional Reputation

1. Either Despine or the printer erred by giving 1834 as the publication date of this work. It is 1824.
2. For more about Micheline Viollet, see Despine's endnote 3, endnote 11, and patient 4 in endnote 15.

Endnote 9: The Impact of Extreme Emotional Experiences

1. Concerning Henriette Marque's death, see Patient 5 in Despine's endnote 15.

Endnote 10: Patients' Distinctive Language

1. See the New Testament, Romans 7: 15–20.

ENDNOTE 11: THE MAGNETIC FORMULAS PATIENTS IDENTIFY

1. In the eighteenth century, theories about electricity in nature—specifically in lightning—were proven by Benjamin Franklin and by the Frenchmen Thomas-François Dalibard and a Monsieur Coiffier who worked with Franklin's theories in Marly, France (Schiffer, Hollenback, and Bell, 2003, pp. 163–164). Despine believed that electrical charges coming naturally from the atmosphere affected the body, just as did electrical charges from a machine or from metals.
2. *Despine's footnote:* Dr. Silvain Aymard, who has worked a good deal with magnetism, wrote:

   G... November 12, 1838

   "You find, my dear colleague, that as regards Mademoiselle Pigeaire, I am treating our *Académie Royale de Médecine* a bit cavalierly and that 'the pear not yet being quite ripe,' you are proceeding more carefully, or (to be more accurate) less frankly than I. But, in truth, my dear colleague, you are mistaken somewhat on the extent of your magnetic knowledge. I assure you I have never yet published anything as extraordinary as what you tell me about Estelle. Her ability to enter into crisis by herself and at will, often to amuse her mother, is quite contrary to accepted knowledge about magnetic action. The fact that you are an enlightened, conscientious observer adds credence to your reports and may be important to quiet critics. I mention only this example, but your publication names many other cases no less prodigious. They prove, unquestionably, that for you and your readers 'the pear is completely ripe.'"
3. *Despine's footnote:* In order for this experiment to succeed, the hot iron must be applied to the experimental circuit just long enough to produce the desired effect, but not longer. Conditions are no longer right for the experiment if the temperature is the same on the whole circuit.
4. *Despine's footnote:* For an experiment on electromagnetic action done by Seebeck (1823), see *Annales de physique et de chimie.*
5. The Latin phrase refers to the healing powers of nature.
6. *Despine's footnote:* Colleagues who make medicine a purely industrial enterprise may accuse me of undermining the field at its foundation by according too much importance to Nature and to somnambulistic patient instinct regarding treatment. These doctors wonder, "Why work for ten years to study science if we must then let patients do as they wish?" This recrimination, whether true or false, is a common complaint among physicians. I respond, "Fear nothing, gentlemen! There are countless illnesses for you to treat and enough human misery, even if you leave to Nature the treatment of nervous disorders (for which you have no interest). Furthermore, the most lucid somnambulist will never say anything, or at least will never say anything

useful, if not well directed and questioned by a learned individual. Hence, the indispensable need to have an above-average doctor treat these patients. An ordinary one, who wants to lead rather than obey Nature or his patient, will make foolish errors."

7. *Despine's footnote:* Carlo Amoretti's (1808) *Della Raddomanzia ossia elettrometria animale.*
8. *Despine's footnote:* This young lady felt an intense burning pain in her elbow or finger when she read with either. Moreover, the finger or elbow swelled and became red, shiny, and tensed, as in erysipelas[g] or a light bout of gout. She is my first patient to suffer from neurological phlogosis.[g] Later, another patient manifested a less pronounced nuance of this phenomenon.
9. Despine is probably referring to cranial nerves that are numbered, not cervical pairs.

Endnote 12: The Magnetic Power of Count Paul D.

1. The state of therapeutic attunement or rapport was practiced and discussed by the Chevalier de Barberin of Lyon, one of the very early magnetizers. He deemphasized the significance of the vital magnetic fluid and stressed the psychological therapeutic relationship with those he sought to help (in some ways rapport can be understood as sympathy, a term Despine used throughout the monograph).

Endnote 13: Letters

1. *Despine's footnote:* Area of Aix where the L'Hardy family spent its first months.
2. Two comments here imply that Estelle and those with her lived at Despine's house in town, not at his country house in Saint Innocent, north of Aix.
3. Currently, this would be the realm of energy psychology.
4. *Despine's footnote:* This patient never came to Aix. I do not know what became of him after Doctor Bottex mentioned him here.
5. *Despine's footnote:* She had arrived in Aix on July 20, 1836 and left on August 3, 1836. She was well enough to think she was completely healed.
6. *Despine's footnote:* For her second stay in Aix, Mademoiselle Augustine Picat arrived on February 2, 1837 and left between April 15 and 20.
7. Adhered to the inside cover of Despine's *Rapports et Notes pour les Bains 1844–1850* is this Magnetizer's Prayer: "Infuse in me, oh God, your divine essence, Holy Spirit, creative breath. And in my weak hands put as much power as charity in my heart" (A. Despine, *Bains d'Aix*).
8. *Despine's footnote:* "Dream" was the name he gave to a charming, short letter Estelle (in somnambulism) had written him.

9. See in the New Testament the Gospel of Matthew 25: 14–30.
10. *Despine's footnote:* Like many cataleptic somnambulists, Estelle always addressed Despine with the familiar form of "you" while she was in crisis. Out of crisis, she always used the polite form of address.
11. Despine's footnote pointing out Estelle's use of the familiar form of "you" in crisis is particularly interesting to the translators who note that Estelle uses the formal form of "you" in Madame L'Hardy's transcriptions of the magnetic sessions. This contradiction is difficult to explain. The translators surmise that Estelle probably did, indeed, use the informal form (as Despine states), and that Madame L'Hardy deliberately chose to substitute the formal, acceptable, and polite mode of discourse for these interactions between a female patient and an elderly male physician. This choice conforms to what we know of the L'Hardy family's strong desire to protect Estelle's reputation.
12. Despine gave a somewhat different account of this lost coin episode of May 7, 1837 in Phase Two of the Magnetic Cure.

Endnote 15: Patients and Related Reports

1. Despine may be referring to detailed drawings of patients Madame Schmitz-Baud, Adèle Maridor, and Sophie La Roche now in the *Archives Nationales des Thermes Nationaux d'Aix-les-Bains.* The artist is unknown.
2. See the item "apparent death" in the Glossary.
3. See Glossary for definition of Second Sight. Despine also attributes this ability to his patients Annette Roux and Henriette Marque.
4. *Despine's footnote*: Annette called the strange sensation she felt in the hypogastrium a "frog." It guided her instinct.
5. See Despine's endnote 8 for Dr. Joseph Marie Souquet's (1769–1839) examination of Micheline in Aix.
6. In some places Despine fails to provide enough data to give the reader a clear understanding of Patient 5 and her narrative. Because some dates cannot be reconciled and we cannot discern Despine's intention, they appear here as they appeared in the French original.
7. For information about an insulated chair, see Despine's endnote 4, note 4.
8. A dictum attributed to Aristotle's *Nicomachean Ethics*, translated as "I love Plato, but I love truth more."

Glossary

aigrette	Unlike the instantaneous spark, the aigrette (also called the electric feather or electric plume) created a sustained electrical discharge. It was administered with a rod-like device with several fairly sharp points at the tips (Lucas, 1900)
alkali	a carbonate or hydroxide of an alkali metal; its aqueous solution is bitter, slippery, caustic, and usually basic in chemical reactions
aphonia	loss of the voice following disease, injury to the vocal cords, or psychological conflicts
apophysis	a natural swelling, projection, or outgrowth of an organ part, such as the process of vertebrae
apoplexy/apoplectic	(1) sudden impairment of neurological function, especially when resulting from a cerebral hemorrhage; (2) a sudden effusion of blood into an organ or tissue; (3) in colloquial French means to "have a fit of temper"
apparent death	this phenomenon may be best understood as a deep trance reaction; in the late Enlightenment, electrical apparatuses were used to resuscitate the apparently dead whose life signs, including pulse, indicated that the person's physical functions had stopped (Schiffer, Hollenbach, and Bell, p. 156)
arnica	(1) any of various perennial herbs of the genus *arnica* in the composite family, having opposite simple leaves and flower heads; (2) a tincture of the dried flower heads from the European species *A. Montana* applied externally to reduce pain and inflammation of bruising and swelling
Athalie	a classic play by Jean Racine (1639–1699) published in 1691
atavistic	used to describe the reappearance of a characteristic after several generations of absence
ataxic	descriptive of ataxia, a loss of muscle coordination
bilious	peevish, irritable, cranky
bouillon	very hot water bath; from the French verb 'bouillir', to boil

brushing	brushing increases circulation; a variety of brushes were used for different purposes
Cabias	Dr. Jean-Baptiste de Cabias (1623) wrote *Les merveilles des bains d'Aix-en-Savoie*
calotte	top of head, skull
cataplasm	poultice
catarrh	inflammation of mucous membranes, especially of the nose and throat
cauterize	to burn tissues, usually to destroy damaged or diseased tissue
cautery	any substance, electric current or hot iron used to destroy tissue
chilblain	(1) inflammation followed by itchy irritation on the hands, feet, or ears, resulting from exposure to moist cold; (2) frostbite
chyle	a milky fluid consisting of lymph and emulsified fat extracted from chyme by the lacteals during digestion and passed to the bloodstream through the thoracic duct
clyster-pipe	an enema
compression	a squeezing together
cornets	cone-shaped cupping glasses (or suction cups) used to create suction in an unhealthy area of the body and bring fresh blood to it
cupping	treatment in which evacuated glass cups are applied to skin to draw blood to the surface; this vacuum is thought to relieve internal congestion
cupping glasses	see cornets
cutaneous	pertaining to the skin
Desbordes-Valmore	Marceline Desbordes-Valmore (1786–1859), a romantic poet
docimasia	(1) a branch of metallurgy, docimasia was the practice of testing minerals for their quality and quantity of metals; (2) a method used to determine whether a dead infant was stillborn by placing the body in water; the body sinks unless the infant has expanded the lungs in respiration
Double View of the Hebrides	see Second Sight
edematous	swollen due to excessive fluid in the body tissues
emprosthontonic	relating to a emprosthotonos—a tenanic spasm when the body flexes forward (opposite of opisthotonic)
encephalon	the brain of a vertebrate

epigastrium	the upper-middle part of the stomach region
erisypelas	(1) superficial bacterial skin infection characteristically extending into cutaneous lymphatics; (2) St. Anthony's fire; a febrile disease accompanied with a diffused inflammation of the skin, which, starting usually from a single point, spreads gradually over its surface. It is usually regarded as contagious, and often occurs epidemically
etiology	(1) the study of causes or origins; (2) the branch of medicine that deals with the causes or origins of disease
febrifuge	agent that reduces a fever; an antipyretic
fistulous	tubular and hollow
frontal bump	frontal bone, forehead
fumigation	to apply smoke or fumes in order to disinfect, perfume, or treat a condition therapeutically
gibbosity	the condition of being gibbos, or hunchback
Gall	Franz Joseph Gall (1758–1828), scientist, phrenologist, anatomist
Garus's Elixir	medicine to treat certain stomach aliments
gastritis	inflammation of the stomach, especially of its mucous membrane
globus hystericus	sensation of a lump in the throat or difficulty swallowing for which no medical cause can be found, hence usually associated with anxiety
goiter	a noncancerous enlargement of the thyroid gland, visible as a swelling at the front of the neck
gros	one gros is one drachma or one-eighth of an ounce
hematuria	blood in the urine
Hippocrates	Hippocrates of Cos (460–380 BC) known as the Father of Medicine
horripilation	the bristling of body hair, or the sensation of creeping flesh (goose bumps) due to cold, fright, or such
humoral	relating to body fluids, especially serum
hypnogogic	(1) the period of drowsiness between wakefulness and sleep; (2) intermediate consciousness before sleep
hypocondrium	upper lateral region of the abdomen, below the ribs
hypogastrium	lowest of the three median regions of the abdomen
inguinal	of, relating to, or located in the groin
innervation	(1) distribution of nerves in a body part, muscle, or organ; (2) stimulation of a nerve, muscle, or body part to action
insufflation	the act of blowing breath on the body, or of blowing air, gas, powder, or vapor into any cavity of the body. Magnetizers thought that their blowing (puffing) on the patient was one means of transmitting the vital magnetic fluid, therefore insufflation had, they believed, a therapeutic impact
Lamartine	well-known romantic poet Alphonse de Lamartine (1790–1869) made eight visits to Aix between 1816 and 1830

laudanum	(1) an alcoholic tincture of opium, used for various medical purposes; (2) coined by Paracelsus for a medicine he mixed, supposed to contain gold and crushed pearls and many expensive ingredients, but probably most effective because it contained opium
lipothymia	(1) loss of consciousness, more or less complete, with loss of motor faculties but maintenance of circulatory and respiratory functions; (2) a condition or feeling of faintness
lockjaw	trismus, see tetanus
lymphatic	having the characteristics of flabbiness, sluggishness
magistral	signifies a medication prepared according to a physician's prescription
meninges	membranes covering brain and spinal cord
metritis	inflammation of the uterus
Millevoie	a melancholy poet whose work typified romantic literary ideals, Charles-Humbert Millevoie (1782–1816), particularly known for his *La chute des feuilles* published in 1812
moxa	(1) use of intense heat at a designated point to increase blood circulation; (2) a cone or cylinder of cotton wool or other combustible material placed on the skin and ignited in order to produce counterirritation
occiput	back part of head
occipital indent	indent on the occipital bone, a curved trapezoid compound bone that forms the lower posterior part of the skull
opisthotonic	of or relating to opisthotonos—tenanic spasm in which the head and heels are bent backward and the body is bowed forward
parietal	pertaining to or forming (1) the wall of a cavity; or (2) the parietal bone, one of two bones that together form the roof and sides of the skull
pericardial	of or relating to the pericardium
pericarditis	inflammation of the pericardia
pericardium/pericardia	the membranous sac enclosing the heart
phlogosis	inflammation of external parts of the body
pneumogastric	pertaining to the lungs and stomach
Pott's Disease	vertebral tuberculosis named after the English surgeon Percival Pott (1714–1788) who studied the disease
prodrome	early symptoms indicating the onset of an attack or disease, a premonitory symptom
purgative	laxative

pyretic	feverish
pyrexia	fever
quinquina	alcoholic bitters with quinine
quintal	100 kilograms
sagittal	(1) of or pertaining to the suture between the parietal bones at the roof of the skull or to the venous canal within the skull and parallel to this suture; (2) in direction or location from front to back in the median plane parallel to the median
sanguine	(1) cheerfully optimistic, hopeful, or confident: a sanguine disposition*; sanguine expectations;* (2) in old physiology having blood as the predominating humor and consequently being ruddy-faced and potentially impulsive and rageful
Second Sight	the ability to see what is happening far away—also called Second Sight of the Hebrides or Double View of the Hebrides
sinciput	(1) the upper half of the cranium, especially the anterior portion above and including the forehead; (2) the forehead
St. Guy's Dance	epilepsy
strumous	(1) having struma or scrofula—a form of tuberculosis affecting the lymph nodes, particularly in the neck; (2) a goiter
superciliary indent	eyebrow
syncope	brief loss of consciousness due to a temporary deficiency of oxygen in the brain
Tarentism	(from Tarente, a southern Italian town) a nervous disorder marked by stupor, melancholy, and uncontrollable dancing mania; popularly attributed to the bite of a tarantula (Venes). Music was one treatment for these symptoms. The word now sometimes refers to music therapy. See the Old Testament, 1 Samuel 16:23
Tetany/tetanic	a set of symptoms or signs (including hyperreflexia, carpopedal spasms, cramps, and laryngospasm) caused by low blood calcium
tetanus	a disease that causes severe intermittent spasms such as in lockjaw
Tigretier	a dancing mania or form of tarentism caused by the bite of a poisonous spider, occurring in Tigré, Abyssinia (Venes)
tisane	a herbal tea
torticollis	spasmatic contraction of the neck muscle during which the head is twisted to one side
transudation	the process of passing through pores or interstices in the manner of perspiration

transposition (or transfer) of the senses	the belief that, in an altered state, a person's sense (for example, of sight) would be relocated from its usual anatomical place (the eyes) to another part of the body (such as the fingertips, wrist, or stomach)
trismus	see lockjaw
tumescence	a swelling or enlargement
Tyrolean	belonging to or relating to Tyrol (or Tirol), a province in the Alpine region of West Austria
vesicatories	blistering agents

Works Cited

Addison, C.G. (1842). *The history of the Knights Templar, the Temple Church, and the Temple.* http://www.templarhistory.com/Chapter9.html.

Anderson, K., L. Anderson, L., and W. Glanze (1994). *Mosby's medical, nursing, and allied health dictionary*, 4th edition. St. Louis, MO: Mosby.

Archives de l'Etat, Neuchâtel.

Archives départementales de Haute-Savoie: Fonds Garbillon-Despine, Sous-série 11 J, Inventaire Robert Gabion, Chargé d'études documentaires, Annecy, 1981; Fonds Aussedat-Despine, Sous-Série 45 J, Inventaire Robert Gabion, chargé d'études documentaires, Annecy, 1983.

Archives des Thermes Nationaux d'Aix-les-Bains: Fonds de la correspondance des médecins Joseph, Antoine et Constant Despine, Père, Fils et Petit-Fils, 1787–1843, Inventaire de monsieur le Professeur Y. Custa, Aix-les-Bains, Rapport au gouvernement, 1994.

Archives privées de Philippe Despine de Dijon, France.

Barach, P.M. and Standards of Practice Committee. (1994). *ISSD Guidelines for treating dissociative identity disorder (multiple personality disorder) in adults.* Skokie, IL: International Society for the Study of Dissociation.

Bonjean, J. (1838). *Analyse chimique des eaux minérales d'Aix-en-Savoie.* Chambéry: Puthod.

Brown, D.P. and E. Fromm (1986). *Hypnotherapy and hypnoanalysis.* Hillsdale, NJ: Lawrence Erlbaum.

Crabtree, A. (1988). *Animal magnetism, early hypnotism, and psychical research, 1766–1925: An annotated bibliography.* White Plains, NY: Kraus International.

Despine, A. *Bains d'Aix: 1844–1850. Rapports et notes par Baron Antoine Despine.* Archives des Thermes Nationaux d'Aix-les-Bains.

——— (1840). *De L'Emploi du magnetisme animal et des eaux minérals dans le traitement des maladies nerveuses, suivi d'une observation très curieuse de guérison de névropathie.* Paris: Germer Baillière.

——— *Manuscrits et notes, Travel Journal, 25 octobre 1841 au 29 janvier 1842.* Archives privées de Philippe Despine de Dijon, France.

Despine, C. (1834). *Manuel de l'étranger aux eaux d'Aix-en-Savoie.* Anneci: Burdet.

Despine, P. (2005, unpublished manuscript). *La Biographie d'Antoine Despine: 1777–1852.* Translator of text: Kimberly Mabry. Translator of footnotes and Editor: Lauren Anderson.

Dictionnaire de médecine (1823). Paris: Bechet Jeune, Librairie de l'Académie royale de médecine.

Dictionnaire historique et biographique de la Suisse (1921). Neuchâtel: Attinger.

Ellenberger, H. (1970). *The discovery of the unconscious: The history and evolution of dynamic psychiatry.* New York: Basic Books.

Fara, P. (2005). *Fatal attraction: Magnetic mysteries of the Enlightenment.* Series: Revolutions in Science. Cambridge and London: Icon.

Finger, S. (2000). *Minds behind the brain: A history of the pioneers and their discoveries.* Oxford: Oxford University Press.

Fink, D. and M. Golinkoff (1990). MPD, borderline personality disorder, and schizophrenia: A comparative study of clinical features. *Dissociation*, 3, 127–134.

Foissac, P. (1833). *Rapports et discussions de l'Académie royale de médecine sur le magnétisme animal, recueillis par un sténographe, et publiés, avec des notes explicatives.* Paris: Baillière.

Kluft, R.P. (1984). An introduction to multiple personality disorder. *Psychiatric Annals,* 14, 19–24.

Lucas, F. (1900). *Electricité médicale: traité théorique et pratique. Précis d'électricité Appareils et instruments électro-médicaux Applications thérapeutiques.* Paris: Librairie Polytechnique Ch. Bérenger.

Mackaman, D.P. (1998). *Leisure settings: Bourgeois culture, medicine, and the spa in modern France.* Chicago: University of Chicago Press.

Magictile(*carreaumagique*).http://electropolis.tm/tr/statique/electrostatique/pages/carreau/html.

Moore, R. (1873). *The Artizan's guide and everybody's assistant: Containing over two thousand new and valuable receipts and tables.* Digitized on November 8, 2005. http://books.google.com/book?id=ThWvM5m4ux8C.

Moreau, P., de Tours. (1888). *De la folie chez les enfants.* Paris: Baillière.

Northcutt, W. (1996). *The regions of France: A reference guide to history and culture.* Westport, CT: Greenwood.

Philippe, J. (1883). *Manuel biographique de la Haute-Savoie et de la Savoie, contenant pour chaque canton une notice sur les principaux personnages qui y sont nés et se sont fait remarqués comme hommes de science, écrivains, militaires, ou ayant rendu des services à leur pays et à leurs concitoyens. Destiné aux écoles et aux établissements d'instruction publique*, Annecy, pp. 19–20.

Republican Guard (*La Garde Républicaine*). (2006). Ministère de la Défense. Historique de la Garde. http://www.garderepublicaine.gendarmerie.defense.gouv.fr/histoire.html.

Robertson, W.C. (2005). *Stop faking it: Finally understanding science so you can teach it.* Arlington, VA: NSTA Press.

Schiffer, M.B., K.L. Hollenbach, and C.L. Bell (2003). *Draw the lightning down: Benjamin Franklin and electrical technology in the Age of*

Enlightenment. Berkeley and Los Angeles, CA: University of California Press.

Socquet, J.M. (1824). *Essai analytique-médical et topographique sur les eaux minérales, gaseuses, acidules et thermo-sulfureuses de Lapperrière près de Moutiers en Savoie*. Lyon, France: Jean Barret.

Turner, G. L'E. (1998). *Scientific instruments: 1500–1900. An introduction*. Berkeley, CA: University of California Press.

Venes, Donald. (1997). *Taber's Cyclopedic Medical Dictionary*. 18th edition. Philadelphia, PA: F.A. Davis, Co.

Name Index

Subject Index

Head Note

In Despine's time, the precursor of hypnotism—here indexed only as mesmerism—was known then by various other names such as animal magnetism, crisis or crisis state, magnetic sleep, somnambulism (artificial, magnetic, spontaneous), or somnambulistic state. All these variant names are indexed under mesmerism.